W9-BEC-191

## Other Titles Available
## from Dorset House Publishing Co.

*Becoming a Technical Leader: An Organic Problem-Solving Approach*
  by Gerald M. Weinberg

*Data Structured Software Maintenance: The Warnier/Orr Approach*
  by David A. Higgins

*Fundamental Concepts of Computer Science: Mathematical Foundations of Programming*
  by Leon S. Levy

*General Principles of Systems Design*
  by Gerald M. Weinberg & Daniela Weinberg

*Peopleware: Productive Projects and Teams*
  by Tom DeMarco & Timothy Lister

*Practical Project Management: Restoring Quality to DP Projects and Systems*
  by Meilir Page-Jones

*Understanding the Professional Programmer*
  by Gerald M. Weinberg

*The Secrets of Consulting: A Guide to Giving & Getting Advice Successfully*
  by Gerald M. Weinberg

*Software Productivity*
  by Harlan D. Mills

*Strategies for Real-Time System Specification*
  by Derek J. Hatley & Imtiaz A. Pirbhai

# RETHINKING
# SYSTEMS
# ANALYSIS
# & DESIGN

# RETHINKING

# SYSTEMS

# ANALYSIS

# & DESIGN

## GERALD M. WEINBERG

DH

**Dorset House Publishing**
**353 West 12th Street**
**New York, NY 10014**

**Library of Congress Cataloging-in-Publication Data**

Weinberg, Gerald M.
    Rethinking systems analysis and design / by Gerald M. Weinberg.
      p.     cm.
    Reprint. Originally published: Boston : Little, Brown, c1982.
    Bibliography: p.
    Includes index.
    ISBN 0-932633-08-0 (pbk.) : $26.50 (est.)
    1. System analysis.   I. Title.
[T57.6.W425  1988]
003—dc19                                88-5083
                                                    CIP

Cover Design: Jeff Faville, Faville Graphics

Copyright © 1988 by Gerald M. Weinberg. Published by Dorset House Publishing Co., Inc., 353 West 12th Street, New York, NY 10014.

Portions of this book appear in *Becoming a Technical Leader, The Secrets of Consulting,* and *General Principles of Systems Design.*

All rights reserved. No part of this publication may be reproduced, stored in a retrieval system, or transmitted, in any form or by any means, electronic, mechanical, photocopying, recording, or otherwise, without prior written permission of the publisher.

Printed in the United States of America

Library of Congress Catalog Number 88-5083
ISBN: 0-932633-08-0

*To* SIR GEOFFREY VICKERS
*for teaching me a new way of thinking*

# Preface

*Systems analysis in the human field is not primarily concerned with solving problems. Its business is the understanding of situations which are relevant to the concerns of human minds.*

Sir Geoffrey Vickers

There seems to be no cure for congenital worrying. I used to worry because nobody knew how to do systems analysis, but now I worry because everybody knows.

And I used to worry because there weren't any good books on systems analysis or design, but now I worry because there are several. Why should *that* worry me? Because the books are so good people will imagine they can learn analysis or design from a book.

In other words, I worry because you can't learn certain things from a book, but I also worry because you can. As a professional author, would I be held responsible if someone fails to learn analysis and design from this book? From other books you can learn to draw a data flow diagram, construct a decision table, or build a queueing model, but this book promises nothing of the sort. As a matter of fact, *if this book succeeds, you will know less of analysis and design than you knew before!*

If you're still with me, you'll understand why I'm worried that you might not want to buy this book. And if you don't buy it, then others won't buy it and my publisher will hate me and I won't be able to feed my dogs and they'll turn vicious and attack me and. . . .

Perhaps you'll buy the book out of kindness to animals—there are some nice animal stories inside that you might read to your children—but I doubt it. So I'd better explain *why* you'll know less after reading this book than you know now.

In the past decades the potential power of information systems over our lives has grown. Our understanding of analysis and design has also grown—*but not as fast.* In short, we are losing the race. Perhaps I shouldn't worry about it, but there are many rumblings on the political horizon. I'm afraid the public isn't going to put up with us much longer.

I'm also worried that the race won't be won simply by going faster or spending more money on the sorts of things we're now doing. The job of analysis and design has not just grown bigger, it's also become *qualitatively different*.

So, if you retain the same narrow view of analysis and design, your knowledge is growing smaller every day, relative to the problems you face. Of course, if you take the ostrich approach, you can imagine that you're keeping up. But if you pull your head out of the sand—by reading this book, for example—you're going to start being a worrier, like me.

This book is *not an alternative* to the growing movement towards more "structured" design and analysis. These new methods are excellent for the parts of the process to which they apply. But as they succeed, we will see more clearly that there are other parts we've been ignoring—parts which we'll then see are our major problems. This book is, therefore, a *supplement* to the more structured processes of analysis and design—an early warning system about the problems we'll have even when we do all the well-structured parts properly.

And those remaining parts are both big and important. As Vickers observes:

> It seems to me obvious that policy making at all levels from the personal to the planetary involves processes which are not merely logical, though they are equally neither random nor arbitrary. And it seems equally obvious that these processes cannot be explicitly and fully described, though they can be recognized, respected, and even influenced. Reluctance to accept these conclusions is, in my view, part of the unhappy legacy which technology has brought to the regulation of human affairs, a legacy derived from the fact that until the recent past its concern has lain with solving problems, where both the problem and the criteria for comparing possible solutions were already given or set by someone else.

I worry because there *are* parts of the analysis/design process that are almost entirely logical, for these are the parts we technologists love to do. Because these parts are well treated by the "structured" methods, we gravitate to them and exclude the remainder—the parts that "cannot be explicitly and fully described."

These parts can be "recognized, respected, and even influenced," which is what this book is trying to do. But I worry because such parts can never be *mastered* in the way decision tables or data flow diagrams can be mastered. Instead, they need constant *re-*

*thinking.* By every analyst and every designer. On every job. Every day.

And of course I worry because thinking is something none of us likes to do every day, least of all analysts and designers. If we think about matters we've put aside as settled, we may turn up worrisome aspects we overlooked the first time. And who but a worrier likes to worry?

But perhaps you're a worrier, like me. You worry that if you *don't* think about some of the less structured processes of systems analysis and design, you'll soon be eaten by starving dogs. At least you'll be out of a job, or in a less rewarding one. In that case, *Rethinking Systems Analysis and Design* might be the cure you need. Taken in conjunction with a dose of the more structured material, it should relieve the torment of worry—and also help you become a more complete analyst or designer.

But will that help relieve "the unhappy legacy which technology has brought to the regulation of human affairs"? Not to worry!

## Acknowledgment
I wish to express my appreciation to Don Gause who co-wrote portions of this book. His assistance was invaluable.

# Contents

# PART
# I

# The New World of
# Systems Analysis and Design

*The Three Ostriches: A Fable (page 23)*

Systems analysis is a new wine in an old bottle. In the early days, a few decades ago, the analyst's job was converting existing systems to new technologies. For that task, the title of "analyst" was reasonably appropriate, but now the situation has changed. Today's analyst still converts and refines existing systems, but increasingly the job is to apply technology to do new things—things that the previous systems never dreamed of.

As the job changes, the analyst's burden grows. When we aren't replacing an existing system, design no longer follows directly from analysis. Two separate jobs become inseparable, for analysis of what exists yields insufficient information to design what will come to exist.

Today there is a new job, but the old names persist. I would prefer to replace the misleading appellation of "systems analyst/designer," but we love our old bottles, even as we delight in our new wines. A new name might force us to rethink systems analysis. Without it, we'll have to think without being forced.

We need new thoughts on what the analyst does—observing, modeling, designing, thinking—and how the analyst becomes a better analyst—education, professional behavior, and personal development.

## Mastering Complexity

We live in an age of unmastered complexity. One symptom of such ages is the popularity of slogans, for slogans are simple, soothing substitutes for thought. Just as a physician diagnoses diseases of the human body by studying symptoms, so can we diagnose the ills of the political body by studying slogans.

### Scale

But when there are so many symptoms from which to choose, how do we begin? One idea is to seek contradictions—two slogans that express a fundamental, and thus not easily resolved, tension. For instance, the economists have an idea called "economy of scale," sometimes reduced to the slogan, "Bigger is better," or "Big is bountiful." For the first half of the twentieth century, in business and industrial production, this idea seems to have been largely unchallenged.

These days, though, we frequently hear a contradictory idea, sometimes expressed as "diseconomy of scale," but more eloquently as "Small is beautiful." Indeed, we sometimes hear two opposite slogans from the same source, like the barfly who observed that, "we'll just have to make a choice between standard of living and quality of life."

How can we resolve such contradictions? Because they come from the same culture, or even the same mind, contradictory slogans must share something that can be used as a clue. But what can "Bigger is better" and "Bigger is worse" possibly have in common?

How about "Bigger is different"? There's a slogan the two sides obviously agree on, so why not start the discussion there? Instead of competing for the greatest decibel rating, the sloganeers might get down to the specifics of how bigger is different from smaller.

Historically it takes some species of disaster to shake sense into our heads. The "Smaller is better" people trek barefoot back to the five-acre vegetable farm, blissful until someone gets a bad case of hookworm. The "Bigger is better" people build refineries larger and larger until one explodes and cannot ever be made to work again.

4

The newspapers (being naturally aligned on the "Bigger is better" side) are never short of stories of "The Back-to-Nature Experiment and Why It Failed." On the other hand, when something big fails, it is usually spectacular enough to sell papers, so we also read about "How the Supertanker Smothered the Beautiful Beach, but the Birds Were Saved by the Audubon Society." In neither case do we get much useful insight from the newspapers. Journalists believe—perhaps correctly—that readers prefer slogans to analysis.

But you can be sure that the owners of the supertanker and the refinery *are* making analyses, though the would-be peasants may not be so inclined. And when the owners make their analyses, they discover that there is much to be learned by asking, "How is bigger different from smaller?"

## The Cost of Not Producing

For instance, a few years ago the people who build refineries and other chemical plants began to notice a trend in failures. Although a bigger plant was more economical when in production, it naturally cost more to have it out of production. Moreover, because it was bigger, there were more things that could go wrong, so things went wrong more often—things that could take it out of production. And finally, because it was bigger, it was generally more difficult to figure out what had gone wrong, so each time it failed, it was out of production for a longer time.

Now, just as you can compute the cost of producing one unit, you can also compute the cost of not producing—that is, the cost added to each unit of production by the failures of your production facility. In Figure 1, we see the two costs plotted versus plant size. The plant could be an oil refinery or a chemical processing plant of any kind. When the plant is small, the production curve is orders of magnitude bigger than the failure curve, so failure is merely a minor nuisance. The plant is simple, so it doesn't fail often, and is easily fixed when it does fail. But even while it is out of action the loss of production is small, so the net effect is usually ignored.

The effect of failure will continue to be ignored as plant size is increased by a factor of ten or even 100. At a certain point, the failures will be noticed by the operating people. But at the higher levels of management there is no recognition of the problem, for

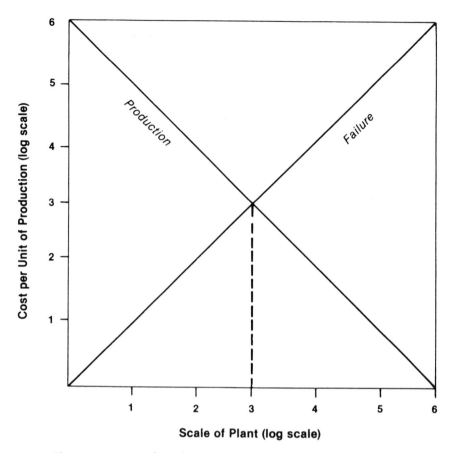

**Figure 1.** Cost of Production and Cost of Failure to Produce

it doesn't affect the overall profit–and–loss statement by any noticeable amount.

As plant size increases once more, upper management may finally become aware of the cost of failure, but will attribute it to "loose operating procedures," or "worker indifference," or "poor quality control." All of these may be factors affecting the slope of the failure curve, but the general nature of the curve has to do only with the ultimate difference between bigness and smallness.

Not understanding the general nature of this relationship, the executives will put a Scale-4 or Scale-5 plant on the drawing boards. They are merely doing what they have always done—the very actions that made them into high executives in the first place.

Not until the plant has been built, failed many times, and been written off, will they seek the analysis that will limit the size of future plants. They will then make up some other slogan, such as "Never more than Scale Four," and begin a new cycle of nonanalysis.

## Mastering Complexity with General Systems Thinking

There are several morals to this story. Most obvious, of course, is that analysis can be cheaper than blind extrapolation. But more careful reading of the story may show how such analysis can be guided by general principles.

For example, although the story comes from chemical engineering, it has nothing to do with the nature of chemicals. The same story has been repeated, or will be repeated, in manufacturing industries of all kinds, including the industry that manufactures information—the computer industry, or information processing industry. Thus we learn that an analyst might learn valuable lessons from outside a particular type of business. This lesson runs counter to the trend toward making the systems analyst a specialist in one particular industry, or even in one particular factor in one particular industry.

Another point: although the story seems to be about chemical plants, or plants of some kind, it's not about plants at all—it's about *people*. It's about how people think, or how they fail to think. It's also about how their thinking combines with their observations and their preconceptions to form their actions. The story tells us about human inertia, which happens to be one of the most general laws governing human behavior, called Weinberg's Law of Twins. (See *On the Design of Stable Systems.*)

The story also tells us, then, that it may be particularly useful for a would-be analyst to study people more than machines. After all, a chemical plant might be very different from an information plant, but one executive is pretty much like another executive. And more like a janitor than like a chemical plant.

Yes, ours is a time of unmastered complexity. Following Weinberg's Law of Twins, we are trying to master this complexity by applying what has worked for us in the past. In particular, we are trying to create a class of specialists to deal with complexity. We call these would-be specialists "systems analysts and designers," though we don't know much about what analysis and design might be.

Well, I'm no exception to Weinberg's Law of Twins, so I have my own ideas about what kinds of things a systems analyst and designer ought to know. As I read the record of the past, what has worked best in dealing with unmastered complexity is a combination of:

1. Learning from analogous situations outside the present situation
2. Learning how people think and combine that thinking with facts and preconceptions to determine action.

The study of these things is what we call *General Systems Thinking,* a subject that I feel forms the core around which the rethinking of systems analysis and design will be built.

# Problems, Solutions, and Systems Analyst/Designers

## Some Mistaken Names

The computer business has always been plagued by egregious misnaming of its fundamental concepts. To take just one example, there is "floating point" and "fixed point" arithmetic. When the decimal point in the computer representation is fixed in its position, we call that the "floating point representation." When the decimal point floats around as a result of arithmetic operations, we call it the "fixed point representation." No wonder students think computer science is a difficult subject.

An even more serious misnaming, I believe, was the choice of the word *analysis* in *systems analysis.* My *American Heritage Dictionary* defines analysis as "the separation of a whole into constituents with a view to its examination and interpretation." Among the many things wrong with this term for describing what a systems analyst does, we might put the following near the top of the list:

1. The analysis only starts when the whole is given—from somewhere undefined.
2. The analysis is not involved with *building* anything, though it may be involved with tearing it apart.
3. To interpret a system by separating it into constituents is a contradiction in terms, at least according to *American Heritage,* which defines a system as "a group of interrelated elements forming a collective entity."

Recently there has been some attempt to remedy this unfortunate choice of terminology by associating the word *design* with the term *systems analysis.* Design means "the arrangement of the parts or details of something according to a plan." *Design,* therefore, adds the active ideas of planning and arranging to the more passive *analysis.* It also, to some extent, emphasizes the *systems* idea—that of parts working together with coherence.

Before we find ourselves adding too many qualifying terms to an initially poor definition, perhaps we would do well to consider a better initial term. It seems to me that a much more appropriate term than *systems analyst* would have been *systems synthesist—* except for the unfortunate difficulty of saying this mouthful out loud. Synthesis is the "combining of separate elements or substances to form a coherent whole," and that seems to say pretty well what we are looking for. Perhaps the term *synthesist* would do all by itself, as *systems* is pretty well implied.

But let me leave the terminology to wiser heads, lest I make the same mistake others have made, and turn instead to defining what it is that a synthesist does or ought to do. Let me begin by defining why a synthesist is not merely a problem solver, as some have suggested.

### The Problem-Solution Myth

> *One could argue indeed that economics has had certain neg-ative effects in that it has improved the efficiency of doing things that probably should not be done at all.*

> Kenneth Boulding

Our dictionary defines a problem as "a question or situation that presents uncertainty, perplexity, or difficulty." While this definition is reasonably clear, our rather precise dictionary begins to founder on the word *solution.*

A solution is "the method or process of solving a problem"—which bases the definition on the word *solve.* We then find that solve means "to find a solution to; answer; explain." This tosses the ball to *answer* and *explain,* yet we find that *answer* means "a solution or result, as to a problem." *Explain* is a little better, for to explain is to "make plain or comprehensible; remove obscurity from."

Looking carefully at these and related definitions, we discern that the "problems," in the sense of "perplexities," may be said to be "solved" if the obscurity is removed. In other words, problems and solutions seem to have an entirely mental existence. Consider, for example, the realm of world affairs, in which we have problems of communism, capitalism, underdeveloped nations, population explosion, or what have you. On the national level we have the farm problem, the civil rights problem, and the problem of medical care for the aged. To each of these "problems," politicians, both professional and armchair, offer their "solutions" and we all jump eagerly into the game. Like Scrabble, however, it is purely a word game, the purpose of which is to "make plain or comprehensible; to remove obscurity from."

Without question, the synthesist will sometimes "solve" problems without building anything more than an edifice of words. More generally, however, when we think of the "systems analyst and designer," we are thinking of someone who may prescribe actual changes to an organization, a community, or a whole society. So the simple view of "solving" problems is inadequate.

But there is an even deeper difficulty with the problem-solver image of the synthesist.

Once we have accepted the reality or importance of problems, accepting the reality or importance of solutions is easy. Ask, therefore, "How is it we began thinking that human life—on any level—could be partitioned into problems?"

We believe that this partition can be traced to the partition of knowledge, and that it is propagated by institutions charged with education and partitioned along various territorial lines.

Consider the following problem: If Bill can chop a pile of wood in five days, and Henry can chop the same size pile in ten days, how long would it take Bill and Henry working together to chop such a pile? Early in our lives we learn that the question of how fast Bill and Henry working together can cut wood is an "algebra problem"—which means that the "solution" is found in an algebra book. We learn, concomitantly, that:

1. The *rate* of cutting wood is the important dimension to woodcutting, since unlike dimensions such as quality of the wood or destruction of the environment, rate of cutting can be *quantified.*
2. The rate of cutting is a quantity *unaffected by the human qualities* of Bill and Henry, such as their inability to work well together, or the inspiration that each gives the other; by *external factors,* such as temperature or other working conditions; or in general by *anything not stated* in the problem.
3. *Everything explicitly stated* in the problem is important, and must be somehow worked into the solution.
4. The solution is a *unique set of symbols* which, when written down, provides the final word, which may be checked by appeal to authority: the teacher or the back of the book.

## The Job of the Synthesist

Each of these four assumptions is violated in my image of what the synthesist does:

1. In systems problems, the important dimensions are given by the human beings involved in the system, either affecting it

or affected by it. The synthesist is not free to choose these dimensions by personal whim, or simply because certain ones are easily quantified. In the end, the synthesist is not allowed to plead innocence of system failure on the grounds that "quality cannot be quantified."

2. Neither can the synthesist plead that "nobody told me that the system would be affected by that." Circumscribing the affecting and affected factors is the first and perhaps foremost job of the synthesist—though little in the university curriculum prepares anyone for that job.

3. If our previous experience with systems analysis proves anything, it proves that anyone who tries to use *all* the information—even about the simple systems existing today—will be drowned in paper and never accomplish anything. The synthesist who doesn't learn early how to ignore 99 percent of everything and get away with it will never make much of a synthesist.

4. Among the 99 percent of everything that can be safely ignored is a mountain of paper solutions to the world's problems. The synthesist is not a politician or philosopher—we already have more than enough of those. The synthesist is someone who makes very specific plans for action, and more often than not stays around during the execution of those plans to adjust them to ongoing reality. And also to do a little wood chopping.

Little wonder, then, that current university education is proving inadequate to the task of turning out future synthesists. Not that I expect a university program to do more than give a good start—a push in the right direction. But at least that would be better than building a set of mythological and terminological barriers and then giving a hard shove backwards.

# The Education of a Systems Analyst/Designer

## What the Synthesist Needs to Know

A *systems analyst/designer*, or perhaps a *synthesist*, is a person who prescribes the building of systems that address human problems. When we consider the scope of systems now under discussion, the burden of this kind of job is staggering. My dictionary gives wide scope to the term *system*, including such diverse alternative definitions as:

1. A network, as for communications, travel, or distribution
2. A set of interrelated ideas, principles, rules, procedures, laws, etc.
3. A social, economic, or political organizational form
4. The state or condition of harmonious, orderly interaction.

The systems analyst/designer we educate today will be contending with all of these definitions tomorrow—and all at the same time in the same system.

Consider, for instance, a communications network for accessing the type of information now stored in libraries. Aside from the technical aspect of communication and storage of information, there is a long list of factors that must be analyzed and designed, such as:

1. The various ideas or images of the system held by different people in the user community—images that influence their behavior in relation to the system and thus influence the operating environment of the system
2. The rules governing the use and abuse of the system, along with the procedures for communicating and enforcing those rules
3. The legal environment in which such a system must operate, including such matters as copyright law, libel and slander, and regulations of interstate commerce, as well as international versions of each of these laws
4. The social structure of the components of the user community, such as the hierarchy in the university, that will determine the distribution of benefits of the system within various user organizations

5. The economics of information—not just within this system, but in relation to alternative systems such as bookstores, public libraries, private collections, special libraries, newspapers, television, face-to-face communication, and competing networks

6. The political realities under which the system must operate, now and in the future, in one country and internationally.

And then, of course, the systems analyst/designer must somehow know how to design a system integrating all these factors (and of course, many more) into harmonious orderly interaction.

Apparently, the education of such a person is going to take time, possibly a lifetime. Possibly even more than a lifetime. And time may be the place where the general systems approach can make its greatest contribution to the education of a systems analyst/designer.

## The General Systems Approach as Learning Amplifier

When I taught general systems thinking in a university environment, I set as one objective of the training the ability to condense learning of new material into short periods of time. Quantitatively, I measured this ability by the time it took a student to learn a new discipline—to the level of a master's candidate in that discipline. I encouraged students to attempt such mastery and to subject themselves to examination by specialists. My better students were able to accomplish this task in three months—in such fields as biology, computer science, statistics, and physics.

Interestingly, the average time seemed longer in such fields as social psychology, political science, and English literature. This may say something about the general systems approach, or it may say something about what is expected of master's candidates in English literature. In any case, the student trained in general systems thinking can move quickly into entirely new areas and begin speaking the language competently within a week or so. The significance of this skill to the systems analyst/designer needs no elaboration.

Of course, general systems thinking is not the only study that amplifies learning. Of utmost importance to the prospective systems analyst/designer are the following studies:

1. Natural language, including linguistics and classical languages to accelerate the learning of technical jargon, plus the mastery of at least one non-native tongue to fluency

2. Mathematics, to the level where the systems analyst/designer cannot be intimidated by clouds of mathematical fog, and can ask intelligent questions of a mathematician.

### Observer Training

Because the systems analyst/designer works with situations that are initially wide open in scope, the ability to observe thoroughly and accurately is essential. The general systems approach concerns itself greatly with issues of what an observer can and cannot know, how observations can be distorted, and how observation can be done most efficiently. In the course of general systems training we use a variety of approaches to training in observation.

Perhaps the foremost technique is total immersion of the student in a problem environment, in order to extract information without prejudgment. By putting several students in comparable environments and comparing the information they extract, we are able to evaluate the quality of observation. Using the paradigms of general systems thinking, we then attribute deficiencies to causes and take steps to remedy them. Our model for this activity is the participant observation of the anthropologist. We might send the student to work in a cannery being considered for automation, to live in a neighborhood scheduled for redevelopment, or to attend classes in a school contemplating some form of computer-aided instruction.

Such immersion demands and trains introspection and attention to detail. Introspection is particularly important because most students are their own worst enemies when it comes to making meaningful observations. Additional training in anthropology, social psychology, and perhaps even personal psychological therapy, are usually important to the prospective systems analyst/designer in learning skills of observation.

Supplementing such field training, the general systems approach makes possible simulations of a variety of observer situations. We use role playing extensively, under the watchful eyes of others in the room or on video tape. We also use a completely abstract "black box" simulation in which students are allowed to stimulate a computer model with inputs and record the outputs. The black box accentuates the interactive nature of the observation process and also removes any claim of "irrationality" in the process being observed.

## Learning to Work with Models

The systems analyst/designer has to work with a variety of models—
mathematical models for analysis of present and future system
behavior, verbal models for communication among the various participants in a system, computer models for both purposes. General
systems training teaches modeling on a level above the specific
forms, concerning itself with the opportunities and traps of models
in general. The general systems student learns to construct models,
verify them, modify them, and relate them to models produced by
other people in other forms.

We want the systems analyst/designer to be able to move
freely among such diversities as mathematical systems theory, classical applied mathematics, computer simulation, physical analog
models, pictorial models of a dozen varieties, and verbal models
in a hundred styles. In particular, our objective is to prevent the
systems analyst/designer from being dominated by one or a few of
these approaches simply because these are the only approaches the
systems analyst/designer knows or feels comfortable with.

Here again, practical experience seems essential. One most
effective exercise is to have the student model the same system
(possibly one on which an observation exercise was done) in at least
three different modes. If it is a class exercise, then within the class
there should be as many different forms of one model as possible—
but also at least two different models of the same form for comparison purposes.

An essential part of the modeling skill is the ability to present
the model effectively, an ability that a good class exercise gives
frequent opportunity to practice under critical eyes. Supplementing
the specific training in modeling should be as much training as
possible in effective writing, speaking, and sketching.

## Becoming a Designer

At our present meager state of knowledge, the weakest part of a
systems analyst/designer training is in design. Part of the problem
is the serious underestimate we make of the amount of effort that
it takes to create a competent designer. No architectural firm would
dream of putting a new graduate to work alone designing a bridge
or an office building, yet these systems are in many ways far simpler
than the information systems the systems analyst/designer is rou-
tinely called upon to design. For one thing, the architect generally

has thousands of successful earlier designs to study and emulate, whereas the systems analyst/designer may have none.

General systems training helps the systems analyst/designer get started on the right foot by providing general principles that apply to all successful systems, or to major categories of systems. These principles substitute to some extent for the nonexistent paradigms from the short past of information systems, but even better, they heighten the usefulness of studying actual paradigms when they are available.

Information systems are large and complex. Without guiding principles, a systems analyst/designer could spend years trying to understand the immense detail in an existing operating system, communication network, or production control system. Even when the system is well documented—and the documentation is available for study—we are easily misled by the narrow view of the original designers. System documentation exists for specific purposes in the life of a system and may not demonstrate what is good—and especially, bad—about the current design. With a set of general principles as a guide, the systems analyst/designer can probe the existing system in a fraction of the time and achieve many times the learning that would be accomplished by simply plunging into a sea of documents.

## Where General Systems Training Fits into a Curriculum

General systems thinking, I believe, is a skill that affects all other learning—hopefully in a favorable way. Therefore it is difficult to isolate training in general systems thinking from any more specific training a prospective systems analyst/designer might receive. The closest analogy that comes to mind is learning about one's native language. Students with strong language skills start any course of study with a substantial advantage over other students less favorably prepared. Mathematics training is another case in point, though perhaps not quite so wide in application.

Consequently, I'd prefer not to see general systems training isolated in a "department of general systems." Instead, I'd like to see general systems thinking taught in every department—and also outside of departments altogether. This seems to be the situation today, if the adoption of my books on general systems thinking is any guide.

But we do have departments of English and departments of mathematics, though the complaining I hear about English language and mathematics skills sometimes makes me wonder. In my brief stay in academe, the thing I learned best was how poorly I understood the academic system. I believe it would be prudent to keep still on the subject of exactly how general systems training ought to be packaged for an academic program for systems analysis and design.

Yes, I think I'll leave it at that. After all, one of the most important and difficult things for a systems analyst/designer to learn is when to keep the mind open and the mouth closed.

### Self-Development

Whatever course the universities take, there will be many skills of importance that they never cover—subjects like learning how to keep your mouth closed and your mind open. That kind of learning is not to be attained in a few years of course taking. For many, it's not attained in a lifetime of experience.

But experience is an excellent route to self-development—particularly if it's the experience of other people. The greatest differences among analyst/designers are not based on differences in schooling or experience, but on what they *do* with their schooling and experience.

The best of the analyst/designers seem to share the ability to apply their own analytical abilities *to themselves*—to think about their own thinking and professional behavior. Can this attitude be taught, or is it simply an intrinsic part of a person's character? Lacking a definitive answer, I prefer to think that we can learn how to learn, and to attain mastery over ourselves. At least we have to try.

This book is my contribution to that effort. I've collected essays of mine from various sources on each of the important topics—general systems thinking, observing, modeling, designing, thinking about thinking, and self-development. I don't believe they cover the entire job of the new analyst/designer, but I do believe they can help anyone who is trying to rethink systems analysis and design.

# Beyond Structured Programming

Before I go any further with the task of rethinking systems analysis and design, I'd like to express myself on the subject of another great "rethinking" in programming—the structured programming revolution. Before anyone gets overly enthusiastic about what the rest of this book says, I want to take stock of what this great rethinking has done. I don't claim to be starting a new revolution of that magnitude, so I'd like people to realize how slow and how small that movement has been so far, in case they think this book is going to make much difference.

My own personal stock taking on the subject of structured programming is based on visits to some hundred installations on four continents over the past few years, plus more than a thousand formal and informal interviews with programmers, analysts, managers, and users during that same period. Because of the conditions under which these visits and interviews took place, I would estimate that the sample is quite heavily biased toward the more progressive data processing organizations. By "progressive," I mean those organizations that are more likely to:

1. Send staff to courses
2. Hire outside consultants, other than in panic mode
3. Encourage staff to belong to professional organizations, and to attend their meetings.

Consequently, my stock taking is likely to be rather optimistic about the scope and quality of the effects of structured programming.

The first conclusion I can draw from my data is that *much less has been done than the press would have you believe.* I interpret the word *press* very loosely, including such sources as:

1. Enthusiastic upper management
2. The trade press
3. The vendors and their advertising agencies
4. The universities, their public relations staffs, and their journals
5. The consulting trade.

Although this may be the most controversial of my observations, it is the most easily verified. All you need do is ask for examples of structured programming—not anecdotes, but actual examples of structured code. If you get any examples at all, you can peruse them for evidence that they follow the "rules" of structured programming. Generally, you will find that:

1.  Five percent can be considered thoroughly structured.
2.  Twenty percent can be considered structured sufficiently to represent an improvement over the average code of 1969.
3.  Fifty percent will show some evidence of some attempt to follow some "structuring rules," but without understanding and with little, if any, success.
4.  Twenty-five percent will show no evidence of influence by *any* ideas about programming from the past twenty years.

Please remember though, that these percentages apply to the code you will actually see in response to your request. If you ask DP installations at random for "structured code examples," about two-thirds will manage to avoid giving you anything. We can merely speculate what their code contains.

The second conclusion from looking at all this material is that: *There are rather many conceptions of what a well-structured program ought to look like, all of which are reasonably equivalent if followed consistently.* The operative clause in that observation seems to be "if followed consistently." Some of these conceptions are marketed in books and/or training courses. Some are purely local to a single installation, or even to one team in an installation. Most are mixtures of some "patented" method and local adaptations.

My third observation is that: *Methods that represent thoughtful adaptations of "patented" and "local" ideas on program structuring are far more likely to be followed consistently.* In other words, programmers seem disinclined to follow a structuring methodology when it is either:

1.  Blind following of "universal rules"
2.  Blind devotion to the concept that anything "not invented here" must be worthless.

I have other observations to make, but now I must pause and relate the effect these observations have on many readers, perhaps including you. I recall a story about a little boy who was playing in the schoolyard rather late one evening. A teacher who had been working late noticed the boy and asked if he knew what time it was.

"I'm not sure," the boy said, "but I know it isn't six o'clock yet."

"And how do you know that?" the teacher asked.

"Because I'm supposed to be home at six, and I'm not home."

When I make my first three observations about structured programming, I get a similar reaction—something like this:

"These can't be right, because if they were right, why would there be so much attention to structured programming?"

In spite of its naive tone, the question deserves answering. The answer can serve as my fourth observation: *Structured programming has received so much attention for the following reasons:*

1. The need is very great for some help in programming.
2. To people who don't understand programming at all, it seems chaotic, so that the term *structured* sounds awfully promising.
3. The approach actually works, when it is successfully applied, so there are many people willing to give testimonials, even though their percentages may not be great.
4. The computer business has always been driven by marketing forces, and marketing forces are paid to be optimistic, and not to distinguish between an idea and its practical realization.

In other words, *structured programming* is precisely like *our latest computer,* in that the terms can be used interchangeably in statements like this:

"If you are having problems in data processing, you can solve them by installing our latest computer. Our latest computer is more cost effective and easier to use. Your people will love our latest computer, although you won't need so many people once our latest computer has been installed. Conversion? No problem! With our latest computer, you'll start to realize savings in a few weeks, at most."

So actually, the whole structured programming pitch was pre-adapted for the ease of data processing professionals, who have always believed that "problems" had "solutions" which could be mechanically applied.

My final observation is related to all of the others: *Those installations and individuals who have successfully realized the promised benefits of structured programming tend to be the ones who don't buy the typical hardware or software pitch, but who listen to the pitch and extract what they decide they need for solving their problems. They do their own thinking, which includes using the thoughts of others, if they're applicable. By and large,*

*they were the most successful DP problem solvers before structured
programming, and are now even more successful.*

There's a lesson in all this that's much bigger than structured
programming or any new hardware or software:

> Our business contains few, if any, easy solutions. Success in
> problem solving comes to those who don't put much faith in
> the latest "magic," but who are willing to try ideas out for
> themselves, even when those ideas are presented in a carnival
> of public relations blather.

Based on this lesson, I'd like to propose a new "programming
religion," a religion based on the following articles of faith:

1.  There's no consistent substitute for a thorough understanding
    of your problem, though sometimes people get lucky.
2.  There's no solution that applies to every problem, and what
    may be the best approach in one circumstance may be pre-
    cisely the worst in another.
3.  There are many useful approaches that work on more than
    one problem, so it pays to become familiar with what has
    worked before.
4.  The trick to problem solving is not just "know-how," but
    "know-when"—which lets you adapt the solution method to
    the problem, and not vice versa.
5.  No matter how much you know how or know when, there
    are some problems that won't yield to present knowledge, and
    some aspects of the problem nobody currently understands,
    so humility is always in order.

I realize that writing a book is not the most humble thing
a person can do, but it's what I do best, and how I earn my living.
I'd be embarrassed if anyone took this book too seriously. We don't
need another "movement" just now, unless it is something analogous to a bowel movement—something to flush our system clean
of waste material that we've accumulated over the years.

# The Three Ostriches: A Fable

Three ostriches had a running argument over the best way for an ostrich to defend himself. Although they were brothers, their mother always said that she couldn't understand how three eggs from the same nest could be so different. The youngest brother practiced biting and kicking incessantly, and held the black belt. He asserted that "the best defense is a good offense." The middle brother, however, lived by the maxim that "he who fights and runs away, lives to fight another day." Through arduous practice, he had become the fastest ostrich in the desert—which, you must admit, is rather fast. The eldest brother, being wiser and more worldly, adopted the typical attitude of mature ostriches: "What you don't know can't hurt you." He was far and away the best head-burier that any ostrich could recall.

One day a feather hunter came to the desert and started robbing ostriches of their precious tail feathers. Now, an ostrich without his tail feather is an ostrich without pride, so most ostriches came to the three brothers for advice on how best to defend their family honor. "You three have practiced self-defense for years," said their spokesman. "You have the know-how to save us, if you will teach it to us." And so each of the three brothers took on a group of followers for instruction in the proper method of self-defense—according to each one's separate gospel.

Eventually, the feather hunter turned up outside the camp of the youngest brother, where he heard the grunts and snorts of all the disciples who were busily practicing kicking and biting. The hunter was on foot, but armed with an enormous club, which he brandished menacingly as the youngest brother went out undaunted to engage him in combat. Yet fearless as he was, the ostrich was no match for the hunter, because the club was much longer than an ostrich's legs or neck. After taking many lumps and bumps, and not getting in a single kick or bite, the ostrich fell exhausted to the ground. The hunter casually plucked his precious tail feather, after which all his disciples gave up without a fight.

When the youngest ostrich told his brothers how his feather had been lost, they both scoffed at him. "Why didn't you run?" demanded the middle one. "A man cannot catch an ostrich."

"If you had put your head in the sand and ruffled your feathers properly," chimed in the eldest, "he would have thought you were a yucca and passed you by."

The next day the hunter left his club at home and went out hunting on a motorcycle. When he discovered the middle brother's

23

training camp, all the ostriches began to run—the brother in the lead. But the motorcycle was much faster, and the hunter simply sped up alongside each ostrich and plucked his tail feather on the run.

That night the other two brothers had the last word. "Why didn't you turn on him and give him a good kick?" asked the youngest. "One solid kick and he would have fallen off that bike and broken his neck."

"No need to be so violent," added the eldest. "With your head buried and your body held low, he would have gone past you so fast he would have thought you were a sand dune."

A few days later, the hunter was out walking without his club when he came upon the eldest brother's camp. "Eyes under!" the leader ordered and was instantly obeyed. The hunter was unable to believe his luck, for all he had to do was walk slowly among the ostriches and pluck an enormous supply of tail feathers.

When the younger brothers heard this story, they felt impelled to remind their supposedly more mature sibling of their advice. "He was unarmed," said the youngest. "One good bite on the neck and you'd never have seen him again."

"And he didn't even have that infernal motorcycle," added the middle brother. "Why, you could have outdistanced him at a half trot."

But the brothers' arguments had no more effect on the eldest than his had had on them, so they all kept practicing their own methods while they patiently grew new tail feathers.

MORAL: *It's not know-how that counts; it's know-when.*

IN OTHER WORDS: No single "approach" will suffice in a complex world, so stay open to new information and don't fall in love with the latest fad.

# PART
# II

# General Systems Thinking

*The Two Philosophers: A Fable (page 40)*

# What Is General Systems Thinking?

*To understand an event, or a sequence of events, means to fit it into a preconceived scheme of thought or perception. It may be a scheme shared by many, or it may be one's own private scheme. At one extreme are the conceptual schemes of science, called theories; at the other, the delusions of psychotics.*

Anatol Rapoport

## The Scientific-Disciplinary-Inductive Approach

*"Facts are the air of scientists. Without them you can never fly."*

Ivan Pavlov

Roughly, the scientific-disciplinary-inductive approach is characterized by a staked-out field of study over which other claimants are not supposed to fly, let alone mine for nuggets. Thus human psychology laid claim to the study of individual human beings, sociology laid claim to groups of human beings, and social-psychology attempted to live on their common border—the study of small groups of human beings.

If a field has found a rich lode, the ranks of its practitioners grow and the claim may be subdivided. Each subdivision then becomes a claim for which the fissioning process may be repeated. Human psychology represents a division of the spoils with animal psychology and is, in turn, subdivided into psychophysics, motivational psychology, industrial psychology, psychology of learning, physiological psychology, psychology of perception, developmental psychology, or what have you.

General systems writers have often deplored this division, but we believe they are mistaken. The method of division is isomorphic to the very methods of science—the understanding of phenomena by analysis. Contrary to the views of these writers, there is nothing inherently wrong with analysis, for many problems yield willingly to this approach.

27

The main trouble with analysis comes when results drawn from one small territory are erroneously applied to another. Practitioners may forget that they are working in one small territory and begin to imagine that their territory is the universe. One of the original goals of The Society for General Systems Research was "To promote the unity of science through improving communication among specialists." We support that goal, and our philosophy of general systems education embodies it.

The general systems approach also begins by staking out a territory—the set of all approaches to dealing with problems. Since the general systems approach is itself part of this territory, we expect that general systems people should be more aware of the limits of their approach than are other disciplinarians. It is not always so.

General systems people are often as parochial as any other students of the universe. When a psychologist is parochial, he at least may be a good psychologist; but when a generalist is parochial, he is—by definition—not being a good generalist. Whatever one may think of psychology, there are undoubtedly good and bad psychologists, and the same is true for general systems people.

The general systems territory is protected against subdivision. If a practitioner begins to subdivide the territory, he ceases for that time to be a general systems person. If he begins to experiment with problem-solving heuristics, he becomes, for that interval, a psychologist. In our view of the general systems approach, such practitioners at such times must be subject to the full judgment of the specialists in each field. There must be no refuge in the claim that they are generalists, not psychologists. If they are doing bad psychology, then they are doing bad general systems work—even though they are not doing general systems work at all.

Clearly, general systems is dependent on scientific-disciplinary approaches, and could not live without the immense body of science. If it is to be a symbiont and not a parasite, it must return something of value to that body.

### The Mathematical-Interdisciplinary-Deductive Approach

*In universities, mathematics is taught mainly to men who are going to teach mathematics to men who are going to teach mathematics to . . . Sometimes, it is true, there is an escape from this treadmill. Archimedes used mathematics to kill Romans, Galileo to improve the Grand Duke of Tuscany's artillery, modern physicists (grown*

*more ambitious) to exterminate the human race. It is usually on*
*this account that the study of mathematics is commended to the*
*general public as worthy of State support.*

Bertrand Russell

When we speak of the "mathematical-interdisciplinary-deductive"
approach, we do not mean mathematics in Russell's first sense,
which is not an approach to dealing with problems but a closed
intellectual pursuit. What we mean is closer to Russell's second
sense, which has indeed all too often been mathematics applied to
the extermination of subsets of the human race. We need only note
with a twinge of sadness that the most prosperous of these branches
of applied mathematics had their origin and greatest success in
military applications—cybernetics and operations research.

Hopefully, these approaches have to some extent outgrown
their military origins, just as they have outgrown their disciplinary
ones. The natural history of such fields seems to begin with specific
problems crying out for solution—problems for which new techniques are developed. Once the new technique has been born it
begins a life of its own, whereupon the discipline also begins independent life.

Although the origin myths of these fields are often recounted
to disprove it, the fields, once mature, start with an approach and
then go searching for a field of application, rather than the reverse.
If scientists can be said to be problem solvers, these interdisciplinarians could be called "solution problemers." Because they start
with a method of solution, their pursuit is not for ever narrower
fields of application, but ever wider ones. As a consequence, their
interdisciplinary activities often find them at cross-purposes with
the disciplinarians.

An analogy would be the proper general systems way to
illuminate the differences, conflicts, and cooperations between disciplinarians and interdisciplinarians. We shall draw upon the ideas
of Owen Lattimore concerning the articulation of sedentary farmers
and nomadic hunters in ancient China, though our interpretation
will be rather fast and loose.

If the disciplinarians are likened to landlords, the interdisciplinarians are like hunters, specialized as to game. Since game
may be found on anybody's territory, hunters must be free to move
about, and thus come to deplore fences. The landlord's worldview,
on the other hand, partitions the world into plots and says that
anything on a plot "belongs" to the landlord. At times, these belief
systems are the source of conflict. Sometimes, however, a hunter

is welcomed onto the holding to stay and hunt for a while, since the landlord may not be a competent hunter, and be overrun with varmints.

If a hunter settles for too long on one territory he runs the risk of becoming a landlord, and we often see this conversion of interdisciplinarians to disciplinarians, just as the Mongol nomad hunters became sedentary Chinese cultivators. General systems people share this danger, but in two directions. In terms of our analogy, the general systems person is a hunter of all game, not just one species. He may, for a time, hunt one particular piece of land: in biology, psychology, or economics, for example. In that case, he becomes for a time a landlord: a biologist, a psychologist, or an economist. On the other hand, he may become fascinated with the pursuit of only one species. In that case he becomes a specialist-hunter—a simulator, a cybernetician, a queueing theorist, or a linear programming expert.

## The General Systems Approach

*At each level there are scientists who apply systems theory in their investigations. They are systems theorists but not necessarily general systems theorists. They are general systems theorists only if they accept the more daring and controversial position that—though every living system and every level is obviously unique—there are important formal identities of large generality across levels.*

James G. Miller

In the preceding section we spoke of the "risk" of being converted into a disciplinarian or interdisciplinarian. This pejorative terminology is common, but we wish absolutely to disavow its use. We do not believe the general systems approach rests on the backs of a professional body of practitioners who are in constant danger of slipping from the straight and narrow path. We believe instead, as Miller implies, that the general systems approach exists without the existence of any full-time "general systems theorists."

To go one step further, we believe that the general systems approach exists *because* full-time general systems theorists do not exist, regardless of current claims. The normal mode in which we practice the general systems approach is to slip into and out of the roles of disciplinarian and interdisciplinarian. Or rather, in Miller's perspective, slipping into the role of general systems theorist, when we are daring enough. As soon as people settle down and begin to specialize in general systems theory, they define themselves out of the business.

Our view is supported by evidence from the careers of the "founders" of the general systems movement. None of them, to this day, lays claim to being a general systems theorist. If you forced Kenneth Boulding to choose his profession (and who could force Kenneth Boulding to do anything?), he would reluctantly admit to being an economist. Ludwig von Bertalanffy was a biologist, or more specifically, a developmental biologist. Miller himself is a psychologist or psychiatrist, the latter profession being shared with Ross Ashby, who also laid claim to being an electrical engineer, or cybernetician, but never a general systems theorist. Anatol Rapoport might volunteer that he is a pianist, or confess to being a game theorist, mathematical biologist, mathematical psychologist, or a political theorist, or a number of other things, but a general systems theorist? Not to our knowledge.

And so it goes through the entire hagiarchy of general systems theory. Which leads us to the first axiom of the general systems approach: Anyone who claims to be a general systems theorist cannot be one, for there is no such thing.

In this sense, general systems is like meditation. There may be people who do research in meditation technique, and there may be people who teach meditation technique, and there may even be people who do both, but there are no people who make a living by meditating. Similarly, there are no people who make a living by "doing" general systems, though there may be people who make a living by saying they are doing general systems.

If our understanding of the general systems approach is correct, it explains, at least in part, why there has been so much confusion. What the disciplinarians and interdisciplinarians share is the existence of practitioners: psychology is what psychologists do and cybernetics is what cyberneticians do. General systems is what we *all* do, when we dare, and yet that which none of us does for a living. Such a beast just doesn't fit within our ordinary thought patterns about what an "approach" should be.

# What Is the System—and Why Does the Question Count?

There's a character on *Sesame Street* called "The Count." He looks like Count Dracula, but his obsession is not with bleeding things to death, but counting them to death. I've never been attacked by a vampire, but I have been attacked by obsessive enumerators. I'm not sure it wouldn't be better to have your blood drained.

Of course we need enumeration if we are to become a "scientific" discipline, but I can't help having nightmares about what can happen when our own Counts start counting. Fear of nightmares keeps me awake, so I try counting sheep. I doze off, and the sheep wander through my dreams. Suddenly I become aware that there is a goat among the sheep. And there's a wolf in sheep's clothing!

Could this nightmare be telling me something? Perhaps it's saying that before you can count sheep, you have to be able to tell the sheep from the goats. And the wolves. One of the most general of the systems principles says:

> BEFORE YOU CAN COUNT ANYTHING,
> YOU'VE GOT TO KNOW SOMETHING

In particular, you've got to be able to place a boundary around the sheep, so that none escapes or enters. Then you've got to be able to tell which are sheep and which are goats.

For instance, we see many figures these days about how much maintenance we're doing, versus how much new development. It seems to me that before we can count maintenance, we have to know how to recognize it. For instance, many "new" development projects are one-for-one replacements of old programs. These development projects could also be considered maintenance, since most of the system will remain unchanged, even though the *program* part is completely replaced.

In other words, if the *system* is taken to be larger than the individual programs, our counting of maintenance versus development work will come out quite differently. From the point of view of the programmer writing the code, the work may seem like new development, but from the point of view of the corporate president, the work is normal, unnoticed maintenance of the functions of the corporation.

In between these two views, there are dozens of other views, each one encapsulating the other into a successively larger system. At the center of the capsule, perhaps, is the hardware. For a long time, we all thought that the hardware was "the system." As late

32

as 1961, I was denounced in public for asserting that programs were systems too, and thus required maintenance. Now the image of the system has grown to include both hardware and programs, but the process shouldn't stop there.

Certainly most people today agree that *job control*—which ties programs together into suites, relates them to sets of data, and creates an interface with the system operators—is part of any system. But a decade ago this concept wasn't understood, and many people failed to count changes to job control as maintenance.

Today, *test data* is often unrecognized as part of the system. As a result, test data and test results are seldom maintained with the devotion appropriate to the cash invested in them. When we develop replacement systems or modifications, the quality of the existing test data base is often the determining factor in success or failure. Yet few designers even consider the quality of this critical asset, because they don't think of it as part of the system.

*Existing files* are another component that is frequently overlooked when counting what is in the system. With a truly pristine application, you have to contend with smudged, dog-eared, illegible filing cards stuffed willy-nilly into rows of dented gray steel cabinets. Converting them to files that can be machine processed makes one humbly aware that existing files *must* be counted as part of the system. Many analysts today have never had to contend with "data in the raw." To them, a gargantuan base of quality data may pass unnoticed. But this data base is also a part of the system, constantly being maintained at gargantuan cost.

Another hidden asset is the vast supply of *documentation* surrounding any decent system. We have operating instructions, user manuals, flow charts, structure diagrams, dictionaries, cross-references, performance measurements, accounting data, and others too numerous to recall.

But even those analysts who remember to include these official forms of documentation in their concept of "the system" are likely to overlook the hundreds of memos, thousands of notes penciled in the margins of manuals, and tens of thousands of scraps of information tucked away in every nook and cranny—like the engineer's phone number written on the partition behind the terminal, or the password taped under the table in the coffee lounge.

If we did an accurate accounting, we'd find that our investment in all this knowledge on paper is far greater than our investment in software or hardware. But greater still—and still less recognized as part of the system—is our investment in the knowledge tucked away in people's heads. When analyst/designers ignore this "brainware," they create "new" systems in which people feel helpless, at least until they have created new brainware to go with the new system. With a broadened concept of the system, and a consequently broadened concept of development, we might see this thrashing period grow shorter and less violent.

But the thrashing can be made worse if the analyst/designer doesn't count another immense portion of the system. The analyst may recognize *explicit* training procedures, yet no matter how important these may be, most training, most of the time, is *implicit*. In any ongoing system, people—just by being themselves—train other people. Moreover, the hardware, software, and data, for better or for worse, train each new person whose life they touch.

When systems are built from scratch, there are no implicit mechanisms in operation to nurse the transition. Designers have

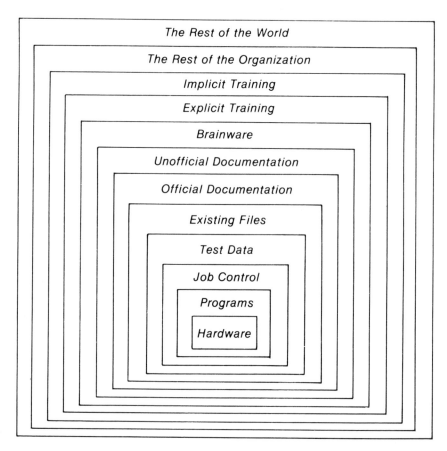

**Figure 2.** The system as nested capsules.

to create explicit training methods—which are always inadequate—and wait until the implicit methods take root.

A designer who is aware that training mechanisms are part of the system can reduce development cost by counting on what exists—by maintaining those valuable parts, rather than sweeping them away. That's the ultimate payoff to a broader understanding of "the system." If we understand that the system is a series of nested capsules, we can accomplish system changes that have exactly the proper continuity with the existing systems that encapsulate them. As suggested in Figure 2, these systems include the even larger organizational context and the rest of the world outside the organization.

If we don't understand this concept of concentric systems, we're likely to design systems that fail when interacting because of some aspect of an "outer" system. We'll alibi our failure by saying, "That part doesn't count—it was outside the system so I didn't have to account for it."

Proper continuity means confessing that every new *development* is actually *maintenance*—of some larger system. But continuity also means we can use these outer systems to make the inner system succeed. Rather than counting them as potential threats to our lifeblood, we can count on them to support our efforts. This kind of counting may be less noble, but it's also less bloody than the Count Dracula method.

# Interdisciplinary Learning

*We build systems like the Wright brothers built airplanes—build the whole thing, push it off a cliff, let it crash, and start over again.*

R. M. Graham

Ed Yourdon begins his book, *Techniques of Program Structure and Design*, with this quotation from *Software Engineering*, the very influential 1969 NATO conference proceedings. I assume that Ed agreed with the sentiments of Professor Graham. I suppose most people in the software business agree with that statement, which is unfortunate, since the statement is utterly false and misleading.

What's false about it? Anybody who has the slightest familiarity with the actual work of Orville and Wilbur Wright would immediately recognize that Professor Graham is graced with a great imagination and a poor store of knowledge. No description could be farther from the truth of the systematic, scientific approach to airplane design taken by the Wrights. More than inventing the first practical airplane, they invented the field of airplane design. They were the first to fly because they, of all their contemporaries, were farthest from the "push-em-off-the-cliff" school of airplane design.

But let's not persecute Professor Graham. When I read his quotation several years ago, I made no particular note of it. I must have accepted it as true and significant. But the year 1978 was the seventy-fifth anniversary of the Wrights' inaugural flight at Kitty Hawk, so during that year I was subjected to a dozen articles and television programs exploring their great achievement. Also, a few years ago I had the good fortune to visit Kitty Hawk. Standing on the beach, next to the commemorative monument, I was so moved my entire body began to shake uncontrollably. I spent hours studying every artifact in the tiny museum, several times being moved to tears. I bought several books that were available, and studied them for months afterwards. Thus, to me, today, Professor Graham's image has become a mockery of an almost sacred history.

Perhaps Professor Graham has also experienced this jubilee year of the Wright brothers, and now regrets his statement as much as I do. If so, both he and I are rare exceptions to the pervasive tendency of computer people to operate in an absolute historical vacuum with respect to any other technology that existed before computers graced the earth.

But we are busy people, building this revolution by pushing systems off the cliffs to see if they can fly. How can we have time for something as frivolous as history? To me, after spending the past years pursuing the history of the technologies of flight, printing, machine tools, telephony, electrical engineering, medicine, and many other fields less obviously related to computing, the question is reversed. How can we *not* have time to study something as relevant as history—especially when we all acknowledge that we who develop systems are in serious trouble? As Gertrude Stein expressed it, "History teaches history teaches."

Just *what* history teaches is quite another matter. As that great American philosopher, Henry Ford, once said, "History is more or less bunk." Statements about the Wrights pushing airplanes off cliffs are just as much part of history as scholarly analyses of their wing-loading calculations. If we study history even for a short while—long enough to read more than one source—we learn perhaps the most important lesson:

There is no one truth about what happened.

Writers lie outright. They make mistakes. They shade or select material to fit their preconceptions or favorite causes. Professional historians are trained to avoid the most egregious errors, but this often serves to make their grand fallacies even more convincing and long-lived. It's dangerous, therefore, to use history for answers to today's problems.

For instance, from an incorrect view of the Wright brothers' methods, we might conclude that we should indeed build systems by pushing them off cliffs (figuratively) and seeing if they fly. After all, we could reason, the Wrights succeeded. Moreover, a few of the more modern methods of aircraft design were notable failures, as in the cases of the Comet and the Electra.

From a more correct view, we might come to the opposite conclusion—that no system should be attempted without at least a decade of calculation and experimentation with models. Although that might have been necessary in order to fly, it certainly hasn't proved necessary for some computer systems.

No, the lessons drawn from the well of history must be more general—more in the line of insights rather than answers. For example, from a study of what happened to the Wrights or other inventors after their creative work was made public, we might be able to envision possible events in our own lives after our creations see the light of day. We might realize, for instance, that invention need not bring happiness or wealth, and that fame itself can be the *worst* result of a creative life.

To take another example of a historical insight, consider the achievement of Gutenberg in the history of printing. My early education had always left me with the impression that Gutenberg's great invention was movable type. In fact, when I just now checked the spelling of his name in my dictionary, I found the following entry:

> Gutenberg, Johann, 1400?–1468?
> German inventor of movable type.

Moreover, I had seen a copy of a Gutenberg Bible with my own eyes—though I wasn't allowed to handle it.

But modern historians demonstrate that Gutenberg was not *the* inventor of movable type, which existed in many parts of Europe in his time, perhaps invented independently. What seems to have been Gutenberg's claim to fame, in his own time, was that he was the first to produce printed works that were sufficiently close to their hand-lettered precursors to be acceptable to the reading and buying public.

It's true that Gutenberg used movable type. He may even have independently invented the concept himself, though this seems unlikely. What he did—what nobody else had done before— was arrange to print large, illuminated letters, in a second color, along with the printing from the movable type. When I saw the Gutenberg Bible I just supposed that those letters had been put in by hand, after the work was printed. That was *my* rewriting of history to fit my own preconceptions. The illuminated letters were not only good enough to fool me, but also to fool Gutenberg's contemporaries. Not fool them, perhaps, but at least allow them to feel comfortable with his printed Bible because it looked enough like a "real" Bible.

Anyone who can't see the relevance of that story to modern computing ought to pause now and discuss it with other programmers or analysts. But the story continues, with many more insights for our age. After Gutenberg was successful at imitating the "real" thing, the public gradually—very gradually—came to accept features in printing that arose from the necessities and economics of movable type. For instance, Gutenberg had to justify right margins—because the monks had done that in lettering by hand. One of the "tricks" the monks used was a collection of equivalent forms of individual letters, some wider and some narrower. As the scribe approached the end of a line, he could estimate the space needed to reach the margin exactly, and then choose which forms of the letters to use to make things come out right.

To match the "real" Bible, Gutenberg had to have many duplicate letters in his set of movable type. When setting them, he had to figure out each line individually. Even though he knew how to justify lines by uniform adjustment of spacing between type elements, his customers would not have accepted such strange looking work.

Over the past 500 years, the extra letters have gradually fallen by the wayside, with a few exceptions in certain type faces for combinations of letters such as *fi* and *ff*. But right justification has remained, because the scribes had done it and because it was easy to do with movable type. Now, however, we computer people are stuck with it, even though it makes extra difficulty for fixed-space printers. Anyone who has ever had to program the uniform sprinkling of blanks in lines of printed text has felt the burden left upon our backs by the monks of old.

Perhaps that's the most significant reason for studying history—to understand where our burdens originated. It may not help to change them, but at least it gives us something to ponder as we strain under them. How many programmers know why numbers are right-justified and words are left-justified, as in preparation for sorting? That particular burden can be understood only in terms of the separate origins of our numeration and writing. When you program in Arabic (as so many of our colleagues do in these oily days), you don't have the problem, because we got our numbers from Arabic scholars. But when you program in Arabic, you have many other problems, because you get your programming languages from the descendants of the Romans.

And when you program in APL, you get a very special combination of right-reading and left-reading—because we got APL from Ken Iverson. But that's another history, all by itself, which perhaps we'll tell another time. It should prove useful, too, for we seldom even study our own history, let alone the history of our technological cousins.

# The Two Philosophers: A Fable

Once upon a time—and come to think of it, not so very long ago at that—two philosophers were out for a ride on horseback before breakfast. Hector, who was riding a large dappled horse named Alice, pointed to the sunrise and said, "Science has made a great contribution to philosophy, for now we know that the sun and the moon are not the same, even though they look alike."

His friend and philosophical opponent, Rector, replied, "You are misled by science, for only philosophy is able to free itself from the illusory 'laws' of the scientists." His horse, a palamino named Fred, tossed his mane in agreement, for Fred was a high-spirited horse.

"How is that?" asked Hector. "Surely you do not believe that the sun and the moon are alike?"

"Of course they are, for both have a luminous essence."

"A luminous essence?"

"Yes, a luminous essence."

"But whatever is a luminous essence?"

"Now, Hector," said Rector impatiently, "don't just make arguments about words. You know perfectly well what a luminous essence is, because it is part of the common experience of all men, and does not depend on language."

"But what if the luminous essence of the sun is not the same as the luminous essence of the moon?"

"Don't just make arguments for argument's sake, Hector. You know that the luminous essence of the two are the same, otherwise you would not call them by the same name."

"You mean, then," said Hector, "that the leg of a horse is like the leg of a table because I call them both legs?"

"Precisely what I mean," said Rector, with a bit of triumph in his voice.

"And therefore you mean that a horse is like a table, because they each have the property of having four legs?"

Rector smiled and reigned his horse. "I do believe you now see the power of my argument. That makes it a good time to stop for breakfast." And so saying, he rode Fred over to the garden restaurant they were passing.

Hector, being hungry, was not inclined to continue the argument so he dismounted and tied Alice to a railing. The serving girl brought them a large loaf of bread and a big iron pitcher of milk.

"Aren't you going to get off your horse?" asked Hector.

"No," replied Rector, "I prefer to eat sitting up high, where I can see the far horizons. Just hand me half of that loaf of bread."

"Still, you should come down and give Fred a rest," said Hector, handing him the bread and going over to take the saddle off Alice.

Fred snorted a little when he saw Alice being unsaddled, but Rector stayed on him anyway. "Be careful where you put that saddle," he told Hector, "the ground is still wet with morning dew."

"Yes, but where else can I put it?"

"Don't be so helpless. Put it on the table."

Hector followed this advice, but the saddle was too big and clumsy for the table, and it knocked off the pitcher of milk. The noise of the iron pitcher frightened Fred, who was already restless, and he reared up on his hind legs and bolted for the road. Rector, who had let go of the reins in order to eat his bread, could not hold on and was thrown to the ground, landing on his head and splitting his skull. Hector merely got his pants and shoes wet with milk, but Rector hurt his head so badly he was not able to philosophize for three months.

MORAL:  *It's true that horses have legs and tables have legs; but it's still better not to eat on horseback or put saddles on the table.*

IN OTHER WORDS:  General systems thinking is fine, but don't ever forget about specific systems thinking.

# PART
## III

# Observation

*The Railroad Paradox (page 56)*

# Can Observation Be Learned in the Classroom?

*Merely looking at the sick is not observing.*

Florence Nightingale

Of all the things we'd like to teach systems analysts, the art of observation seems the most elusive. Perhaps the problem lies in the difficulty of observing the very act of observation itself? How do we know, looking at someone, whether they are observing or not? We all know that some of the most valuable observation skills concern observing without being observed—like learning to read upside down.

In the classroom we have a hard time finding out if the students are even listening to the lecture, let alone noting the important parts. Students may not have learned to observe without being observed, but they've all learned the opposite trick of seeming to observe when not paying attention. Like sleeping with your eyes open.

Because professors don't know when the students are asleep, they often devise tricks to catch them unaware. One notorious law professor stages an interruption to his evidence course in which one man chases another through the lecture hall, threatening to shoot him. When the commotion ends, the professor asks the students to describe the men and the weapon. Many different guns are described, but the actual "weapon" is a banana!

One problem with this method is that few of the students actually believe the professor when he says the "gun" was a banana. In order to teach observation convincingly, you must be able to check the observations against reality. Showing that observers differ is useful, but it doesn't teach the *kinds* of discrepancies that stand between observation and the system observed.

We often use video tape replays to give student observers the kind of solid feedback they need. Video tapes can also be used with staged performances designed to test particular observation skills. The same tape may be viewed several times, each time alerting the observers to a different kind of information. This is the method we use to review training films, with each pass being an inspection

45

keyed to a different observation. On each pass, we "see" different kinds of errors, which helps improve both the films and our powers of observation.

But as valuable as video tape instruction may be, it lacks one essential dimension of observation—the *interaction* of the observer with the system being observed. Possibly the single most common fallacy about observation is that we can somehow separate observer from observed—that the act of observation doesn't influence the observer. If each observer interacts with the system differently, how can we reproduce the experience for purposes of feedback?

For many years now, we have succeeded in teaching systems analysts and others to observe by using a method that allows both interaction and reproducibility. The system observed is reproducible because it is *programmed*—it can be started over from any point in its behavior, and its behavioral model can be examined explicitly in the program. This method is well known in engineering, where it is called "the black box method."

## The Black Box Method

> *The problem of the black box arose in electrical engineering. The engineer is given a sealed box that has terminals for input, to which he may bring any voltages, shocks, or other disturbances he pleases, and terminals for output, from which he may observe what he can. He is to deduce what he can of its contents.*

> W. Ross Ashby

We started with an electric box such as Ashby describes, then graduated to a programmed box used through a batch operating system. Eventually we put the "box" on a time-sharing system, and now on a small, portable computer. With each successive refinement in technology, we have more closely approached our goal of modeling observation in its essence.

By stripping away accidental or incidental features of observation, we have succeeded in training analysts in the deepest and most general principles of observation. We are convinced that exposure to the black box experience in some form or another ought to be an essential part of any systems analyst's training, or retraining.

Over the years we've experimented with many different models "inside" the black box. One of the most important lessons I've learned from these experiments is how difficult it is to learn about the simplest models by this method. Our original models were very complex, which tended to obscure the lessons about observation. Now I realize that an almost trivial set of models will do quite nicely, which makes programming the box rather easy.

For the benefit of anyone who'd like to try this method, I've supplied a detailed description of one set of models in an appendix. The appendix describes the individual boxes that make up a system of black boxes, along with results that students obtain when observing them.

## Learning with the Black Box

*The image has a certain dimension, or quality, of certainty or uncertainty, probability or improbability, clarity or vagueness. Our image of the world is not uniformly certain, uniformly probable, or uniformly clear. Messages, therefore, may have the effect not only of adding to or reorganizing the image. They may also have the effect of clarifying it, that is, of making something which previously was regarded as less certain more certain, or something which was previously seen in a vague way, clearer. Messages may also have the contrary effect.*

Kenneth Boulding

The success of the box, of course, is not measured by information such as that given in Figure D in the appendix showing how many students "solved" each subsystem. Instead, the success must be measured by how much they learn about the act of observation.

The following statements may be taken as typical of the students' verbal and written reports:

1. "Even with relatively simple systems we did not succeed in completely understanding all of them. There are two main reasons for this failure:
   a. Our own limited perception
   b. Preconception or bias concerning what sort of relationship might exist between input and output as well as what aspects of the output may be relevant in the study."
2. "It is extremely difficult to lay out a comprehensive plan which, when executed, will provide all the data required to reach conclusive results. When dealing with the unknown, it appears to be better to handle one situation at a time, examine the results obtained, and then prepare a plan for the next situation."

3. "Boundaries of the system being studied must be carefully defined."
4. "Different systems require different ways of testing."
5. "We found that a team approach to a problem of this nature provided many good ideas which saved time and which one person working alone might have spent many hours trying to localize. In retrospect, however, we realized that we should have organized our computer runs so that each person concentrated on a different aspect of the problem and ran tests individually rather than all three of us working with one deck of cards."
6. "Sheer volume of data is itself an obstacle to analyzing a system."
7. "Effective observational data must carry a pattern designed to uncover complementary patterns inherent in the system under study."
8. "During the first few days teams exchanged information freely. During this period most people were in a state of confusion. However, as results became more conclusive, there was a growing reluctance to exchange information. Teams obviously felt in competition with one another, although no competition was specified in the task."
9. "Different observers looking at the same output do not necessarily reach the same conclusions."

Comments such as these, which could be multiplied tenfold, suggest the richness of the learning experience for the students. To be sure, many of the comments made would be regarded as self-evident by mature systems analysts. Indeed, the students themselves would regard these observations as obvious—if they hear them before they try working with the black box. What they are trying to express in their reports, however, are not platitudes copied from some test or lecture notes, but real insights which they *experienced.*

## Applications to Real Life

*In our daily lives we are confronted at every turn with systems whose internal mechanisms are not fully open to inspection, and which must be treated by the methods appropriate to the Black Box.*

W. Ross Ashby

Not everything about the black box is contained in the individual boxes, for the boxes taken together form a larger system. For instance, there is the behavior of the system when illegal inputs are attempted. Although the system gives explicit rules for what inputs are acceptable, students will always test these rules, either intentionally or by accident. They learn that the written rules are simply another way of observing the system, one which doesn't always give them the whole truth and nothing but the truth.

They also tend to define what is "legal" rather more narrowly than the system allows, thus cutting themselves off from useful observations near the extremes. As one team put it, "Our early encounters with resistance to illegal inputs discouraged us from trying some extreme observations that would have really helped. It's just like on the job. We get punished a little for overstepping some boundary, then we stay *very* far from that boundary in our later studies—just to be sure!"

Another important lesson arises because of the feature that makes the numerical parameters of several boxes depend on the team number. This leads different teams to induce different versions of the transformations, but they generally do not suspect that their version is different from the others.

Class discussion often proceeds for a long time before someone realizes that they are talking about different systems: "What looks like agreement among observers is often an illusion based on poor communication."

When they do realize that their views have been different, they get a second lesson: "What we thought was a general solution was actually a very special case caused by our limited point of view and our unwillingness to extend that point of view by talking with other people who knew things about the system."

Some systems lessons emerge which are not in any sense built into the system. Commonly, a team will imagine that they have found a simple rule when the actual rule is much more complex. They make this error when they constrain their input, unknowingly, to a subset of possible cases. Then their rule turns out to be a rule about their observation strategy, rather than about the transformation. (An example of this kind appears in Figure B in the appendix, if you'd like to experience it.)

One team made the following observation: "We were so confident we understood the system entirely, but it turned out to be a perfect illustration of *The Principle of Indeterminability*. ('We cannot with certainty attribute observed constraint either to system or environment.') We'd all read that in *An Introduction to General Systems Thinking*, but now we really *understand* it."

Another team that had the same experience said, "Now we know why you have to arrange for observations that might *disprove* your hypothesis, rather than trying to prove it."

These are strong and important lessons, but do they last? Because the black box has been used now for almost a generation, there have been many opportunities to discuss its long-term effects. Although we have no quantitative measures, one strong indication of the power of the black box is the number of people who remember the experience after a decade or more. Many former students have testified that the experience changed their personal image of what a systems analyst is attempting to do, and how an analyst might go about doing it better. In short, it started them observing their own selves as observers, and triggered the process of rethinking systems analysis.

### Can Observation be Learned in the Classroom?

*Isn't it splendid to think of all the things there are to find out about? It just makes me feel glad to be alive—it's such an interesting world. It wouldn't be half so interesting if we knew all about everything, would it? There'd be no scope for imagination then, would there?*

Lucy Montgomery
*Anne of Green Gables*

In view of our experiences with video tape, and particularly with the black box, the answer to the title question seems a resounding, "Yes, observation can be learned in the classroom!" But if it's so resounding, why have classrooms so consistently turned out analysts who are poor observers—who must learn these lessons in the excruciating and expensive school of hard knocks?

I believe that my experience with the black box has given me some insight into that question, in a roundabout way. I've always followed the practice of concealing within the black box system a full and explicit description of the individual models, much like the one in Figure C in the appendix. In order to find this description, the students would have to choose just the right combination of "illegal" inputs, which is quite unlikely. In fact, it's never happened.

When I tell them about this "magic key," they groan and cry "Unfair!" This gives me the opportunity to tell them several lessons that all observers must know:

1. There's no reason to expect that a system will be fair to observers.
2. There may be a gold mine of information lying about, if only they know the language in which to ask.
3. Systems often intentionally conceal information from would-be observers, because it may not serve the system's purpose to have too much known about it.

*The most interesting thing about these three lessons is that each is the converse of a lesson taught repeatedly by ordinary schooling!*

We learn in school that teachers are kind, have our best interests at heart, and will always be fair to the students. If they don't do those things, they are bad teachers. We also learn in school that it's not a good idea to ask too many questions. It's good to ask one—possibly two—but then the teacher starts to look at you funny, and you'd better put your hand down. Probing a system past some surface scratching simply isn't polite, and school teaches us to be nice little girls and boys. If there were something we should know, the teacher would tell us about it eventually. And then we would believe the teacher, who must always be right, even when not in agreement with our own observations.

This idea that "the teacher is always right" may be the most destructive of all ordinary classroom lessons about observations. This easily translates into real-life ideas, such as, "If somebody in authority tells me something about this system, then I'd better believe it".

In the black box exercise, I'm invariably surprised by something that emerges from the observations. As a teacher, I suppose I should hide this surprise from the students, lest they should discover that I'm not omniscient. But if I did try to hide my ignorance, they would lose one of the most important lessons of all: "We could see that just because you had built the black box, you didn't know everything about it. So even if we had discovered the magic key that gives us your solution, we shouldn't have believed it—at least without doing some checking on our own."

Perhaps this example contains the answer to our question. Many trappings of the ordinary school are designed to hide various realities from the students, such as the idea that the teacher may not know all the answers, or that there is some essential unfairness in any institution.

We might truly say that all teaching is based on illusion and half-truth. Perhaps this sounds judgmental, but that's only because this truth, too, has been kept from us as students. There's nothing wrong with this kind of concealment, if it serves a good educational purpose. Without restricting the students' environment, we could easily overwhelm them with data they could not possibly process. So good teaching generally revolves around knowing what to leave out, what to conceal.

But for teaching *observation*, the teacher must ultimately be able—and willing—to open up the black box and reveal the white box inside, even if our own personal white box is not quite as perfect as we'd like the world to believe. If we teachers pretend to be what we aren't, we shouldn't be surprised if we teach our students to observe the False and label it the True.

# The Natural History of White Bread

*In its simplest form, learning is the process by which inputs of information in the past lead to images of the future in the present.*

Kenneth Boulding

The request for development has just been approved, so in celebration, the manager takes the analyst to lunch in the executive dining room. They are talking about the forthcoming systems study when the waiter sets a basket of bread on the linen tablecloth.

"Ugh," grunts the manager, "white bread!"

"I'd always thought that executives would eat something better than this raw dough," the analyst complains while buttering a slice with his silver-plated butter knife.

"Now you know it's not such a fancy life, perhaps you'll be happier sticking to your systems studies and leaving the management to me."

"Uh huh," says the analyst, his words muffled by the glop of bread mushed in his mouth. "Let's get back to the systems study. The current system is so poorly designed that I don't know where to look first."

The manager seemed to be having a little trouble swallowing his bread. "Oh, I don't know," he said, "I'm rather proud of the way my design for that system has held up over all these years."

"*Your* design?"

"Yes, didn't you know?"

The analyst tried to cover the embarrassment with an extra thick layer of butter on the next slice of bread. After a long pause, he managed to blurt out, "Well, I guess it might be good to start by studying the history of the system."

The manager smiled wryly. "I think you already have."

Young children lack a sense of history. They imagine that the world that reflects itself into their bright eyes has always and forever existed the way it exists today. Or perhaps it just popped magically into existence the moment before they gazed on it, the way the waiter pops the bread on the table.

Young analysts often take a similar view. Although this solipsistic assumption certainly lends a necessary freshness to their ideas, systems do not live by freshness alone. Those who place excessive value on freshness to the exclusion of other attributes ought to be condemned to a few weeks on water and standard American white bread.

The story of white bread starts when Mrs. Smith, who is poor, starts to bake bread at home to save a little money and get better nutrition. One day, a visiting neighbor remarks on the outstanding quality of Mrs. Smith's bread, whereupon Mrs. Smith offers her a loaf. The first loaf is a gift, but the neighbor's family likes the bread, and soon arranges with Mrs. Smith to bake them a few loaves every week, for money Mrs. Smith can sorely use. Gradually the word diffuses through the neighborhood, and soon Mrs. Smith's oven is working to capacity.

For Christmas Mr. Smith buys Mrs. Smith a new double oven and the baking expands. Soon the children are no longer an adequate labor supply, so Mrs. Smith hires her first paid helper. Her local supplier is unable to keep up with her demands for the high quality ingredients she uses, so Mrs. Smith converts to a commercial grade—cheaper, and available more regularly in large quantities.

And so it goes. As her fame grows far and wide, Mrs. Smith buys her first truck, putting her into the bread distributing business as well as the personnel, purchasing, accounts receivable, and baking businesses. Because of the distribution requirements, bread must be underbaked a day in advance and doped with preservatives to keep it "fresh" on the shelves between deliveries.

One day, Mrs. Smith finds herself the rich owner of the Mrs. Smith Baking Corporation, with her bread gracing the tables of executive dining rooms throughout the nation. But somehow in the process of changing from Mrs. Smith to Mrs. Smith, Inc., the bread has changed from Mrs. Smith's bread to standard American white bread. It's just like all the others because, as Boulding says,

"THINGS ARE THE WAY THEY ARE
BECAUSE THEY GOT THAT WAY."

Somewhere today, a Mrs. Jones has just started another cycle. If only we could persuade Mrs. Jones to study the history of Mrs. Smith, rather than just taste her bread, she might achieve a different outcome. She might avoid mistakes. She might detect small but important changes that Mrs. Smith overlooked. She might keep what worked and change what came out poorly.

But the most important thing Mrs. Jones can learn from Mrs. Smith is that Mrs. Smith failed to study history. And what might history teach the analyst, when a new system study is about to begin? The analyst can learn, with Mrs. Jones, to avoid mistakes, capture missed opportunities, keep what worked, and change what failed. The analyst can learn about the environment too, for even though the system is to be changed, it will have to survive in the environment experienced by the former system.

But even if the analyst learns nothing of direct relevance for the new system, there may be important political reasons to begin with a historical survey. People who produced and participated in the existing system are still around and will be—in one way or another—involved in the development of the new system. They are part of the system's past, but they will also be part of the new system's future, and thus must be reckoned with.

One of the most common errors of young, impetuous analysts is loudly to castigate the developers of the existing system, only to discover that:

1. There were, at the time, good and sufficient reasons for decisions that seem idiotic today.
2. The original developer is now the analyst's manager, or manager's manager.

If the analyst is to avoid such political bloopers, as well as many design flaws, a historical survey is required. Yet not just any survey will do. There remains the matter of the spirit in which the survey is conducted. In order to avoid the pattern of placing blame, the analyst should follow this rule:

STUDY FOR UNDERSTANDING, NOT FOR CRITICISM

Quite naturally, people who built a system now being considered for replacement or modification will be sensitive, since any proposed change could be interpreted as a slur on their intelligence, or at least foresight. For political reasons, the analyst may have to go even further than the above motto suggests, and

SEARCH FOR WHAT IS GOOD IN THE OLD SYSTEM

Besides, what is bad will come to the fore easily enough, if only when we start building our new solution to the problem. Do you think Mrs. Smith doesn't know how bad her white bread is today? After all, she's had her own history to study all these years. Just as your manager has. Just as you have.

And, of course, you study your own history—don't you?

# The Railroad Paradox

About thirty miles from Gotham City lay the commuter community of Suburbantown. Each morning, thousands of Suburbanites took the Central Railroad to work in Gotham City. Each evening, Central Railroad returned them to their waiting spouses, children, and dogs.

Suburbantown was a wealthy suburb, and many of the spouses liked to leave the children and dogs and spend an evening in Gotham City with their mates. They preferred to precede their evening of dinner and theater with browsing among Gotham City's lush markets. But there was a problem. To allow time for proper shopping, a Suburbanite would have to depart for Gotham City at 2:30 or 3:00 in the afternoon. At that hour, no Central Railroad train stopped in Suburbantown.

Some Suburbanites noted that a Central train did pass through their station at 2:30, but did not stop. They decided to petition the railroad, asking that the train be scheduled to stop at Suburbantown. They readily found supporters in their door-to-door canvass. When the petition was mailed, it contained 253 signatures. About three weeks later, the petition committee received the following letter from the Central Railroad:

Dear Committee,

Thank you for your continuing interest in Central Railroad operations. We take seriously our commitment to providing responsive service to all the people living along our routes, and greatly appreciate feedback on all aspects of our business.

In response to your petition, our customer service representative visited the Suburbantown station on three separate days, each time at 2:30 in the afternoon. Although he observed with great care, *on none of the three occasions were there any passengers waiting for a southbound train.*

We can only conclude that there is no real demand for a southbound stop at 2:30, and must therefore regretfully decline your petition.

Yours sincerely,

Customer Service Agent
Central Railroad

This is a true story. If you find it hard to believe, try these stories from a bit closer to home:

1.  A systems analyst in a consumer products company heard that some marketing representatives housed in an auxiliary building might need terminals to access the new marketing information data base. He circulated a questionnaire, which included the question:

    "How much use do you presently make of the marketing data base?"

    People in the auxiliary building, having no terminals unless they cared to walk six blocks, showed zero usage. The analyst concluded that no terminals were required in the auxiliary building.

2.  A systems analyst in a large university was preparing the allocations of "computing dollars" to the departments for the coming year. The anthropology department was asking for $10,000 of computer time, but during the past year they had used no computing time at all, even though their budget was $400. The analyst raised their budget to $500, though he really wanted to cut it, as the anthropologists had not used the computer at all in the past. Of course, the project they had in mind would cost a minimum of $8,000 just to get started, but they had nothing to spend on computing unless it was in the budget.

3.  A systems analyst in a brokerage firm received a request to change the algorithm by which stock movements were forecast. She surveyed the individual brokers, asking:

    "How frequently do you use the stock movements forecast?"

    When they all replied that they never used the forecast, she turned down the request. Of course, the reason the forecast wasn't used was that there was an error in the current algorithm, rendering the output worse than useless.

4.  Engineers at a computer manufacturing company were asked to improve the new version of the company's CPU by adding an efficient mechanism for subroutine calls. After a two-month delay, the engineers responded that they had studied a sample of existing programs and found that hardly any of them used subroutines in situations where efficiency was required. Therefore, they said, the request was frivolous, and would be denied.

Perhaps you now see what is meant by "The Railroad Paradox." Perhaps you would like to add some examples out of your own experience with systems analysis. They aren't difficult to find—when other people are the perpetrators. The real trick is to catch *yourself* acting in the role of "customer service agent"—an agent whose relationship to customer service is the same as that of the county rat agent to the county's rats.

Baldly stated, the Railroad Paradox is this:

1.   Service is not satisfactory.
2.   Because of 1, customers don't use, or underuse, the present service.
3.   Also because of 1, customers request better service.
4.   Because of 2, the systems analyst denies the request, 3.

In short, *because the service is bad, the analyst denies the request for better service.*

If we allow ourselves to be bound hand and foot by the Railroad Paradox, our systems aren't ever going to get much better. The problem is sometimes solved by the customers getting sufficiently violent with the systems analyst. But isn't there some way we can save our heads and our jobs?

Here are some ideas for an analyst who wouldn't like to be a rat agent:

1.   Be aware of the Railroad Paradox. Be sure you understand its unbreakable logic.
2.   Never use customer questionnaires without a certain amount of open-ended discussion with a few of the customers to interpret the reasoning behind their answers.
3.   Use present levels only as a guide, never as a standard, for future performance.
4.   If possible, supply some kind of trial of the new service before deciding to discard it on the basis of poor present use of the old service.

A trial is the most effective test, but does have problems itself. If a service is new, and the old service was bad or nonexistent, a long start-up period may be required. Also, in some systems, users may be wary of getting "hooked" by a trial that they can't rely on being continued. For instance, if they had to change software to use a new compiler feature, they wouldn't make the investment unless they were assured that the feature was going to remain in the standard compiler.

Quite often though, the trial can be made at low cost and risk to both parties. Central Railroad could have tried a stop at Suburbantown for a month or two with little difficulty. The consumer products company could have provided dial-up terminal service to the auxiliary building for a limited time. The anthropology department could have been given a special supplemental budget for one year, on condition that they would have to use it or lose it. The brokerage could have provided a manually computed stock movements forecast for a few weeks with little commitment.

Of the cases cited above, only the argument against efficient subroutine linkage might be difficult to give a genuine trial. The cost might be great, and programmer behavior might be hard to change. Perhaps that's why this particular piece of railroading has been going on for more than twenty years!

# The Dog Who Read Fables: A Fable

A black poodle who was very intelligent—and knew it—was coming home one evening from the public library, where he had just been reading fables by Aesop. He was rather upset with the way Aesop often presented dogs—as not very bright, all things considered. "That story about the dog with the bone dropping it in the water in order to get its reflection," he stormed to himself, "is particularly infuriating. No dog is that stupid—even though one must admit that some are a bit greedy at times."

As he mulled the story over and over in his mind, growing ever more furious with each repetition, he approached a small bridge. In the water under the bridge, there happened to be a large ham bone, full of meat, which had fallen out of the grocery boy's delivery basket. Being a dog, he could not help but notice the bone out of the corner of his eye as he crossed the bridge, in spite of his preoccupations. He stopped for only the briefest instant—in just such a position that the bone appeared to be in the mouth of his reflection. "Aha," he thought, "there's the very same situation happening to me," for indeed he was carrying a small, rather bare, bone in his mouth. "Of course the bone down there in the reflection looks bigger and better than the one I have, but *I'm* certainly not as stupid as Aesop imagined!" And with that comforting thought, he turned away from the temptation and trotted off to a rather meager supper on his tiny bone.

MORAL: *Not all illusions are illusions.*

IN OTHER WORDS: Don't let your desires influence your observations, especially your desire not to appear stupid.

# PART
# IV

# Interviewing

*The Fairy and the Pig: A Fable (page 85)*

## A Surefire Question

Analysts frequently ask me for a list of surefire questions to use in the first interview with a new client. The awkwardness of this first interview seems to be a universal experience among analysts, much like the first date among adolescents. Their request for a surefire list reminds me of the fourteen-year-old boy, about to go on his first date, who asks his older sister for a list of topics that girls like to talk about.

"There are three surefire topics you can use whenever you're lost for something to say. You can ask her about her preferences in food, talk about her family, or bring up a question of philosophy. Can you remember that?"

"Sure," he said, repeating the magic formula to reassure himself. "Food, family, philosophy."

After taking his date to the movies and to McDonald's, he found himself sitting with her on the front porch, unable to vibrate his vocal cords. Then he remembered the first item on his sister's list, and managed to blurt out, "Do you like spaghetti?"

"No."

He almost panicked at the failure of his first question, then remembered the second topic. He stammered, "Do you have a brother?"

"No."

He almost choked, but his mind raced down to the third item on the list, philosophy. What to ask? That's it, he thought, and just managed to utter, "If you did have a brother, would he like spaghetti?"

Not every date presents such problems of noncommunication, but every first interview between analyst and client does. At least you'll be safer if you assume that it does. You may exchange information freely on many topics, but what about the things you forgot to ask about at all? Or the things you asked that weren't relevant to the problem, even though they may have seemed relevant at the time?

In Ursula Le Guin's prize-winning science fiction novel, *The Left Hand of Darkness*, the visitor from another planet is trying to understand the Foretelling practice on the planet Winter, and is discussing it with Faxe, the Weaver.

"Faxe, I don't think I understand."
"Well, we come here to the Fastnesses mostly to learn what questions not to ask."
"But you're the Answerers!"
"You don't see yet, Genry, why we perfected and practiced Foretelling?"
"No—"
"To exhibit the perfect uselessness of knowing the answer to the wrong question."

It's no accident that Le Guin writes about such problems, for she is the daughter of the famous anthropologist, Alfred Kroeber. Anthropologists have a lot in common with systems analysts, for both of them have to enter unfamiliar worlds and extract reliable information. Because my wife is an anthropologist, people often ask me, "What is the difference between an anthropologist and a sociologist?" Many people are familiar with sociologists, perhaps having been interviewed by them at some time in their lives, but few people have ever seen an anthropologist at work. I usually explain that anthropologists work in *other* people's cultures, but that leaves much unanswered. Why, people ask, should interviewing techniques be different in another culture?

I recently learned a good reply to this question. A sociologist knows the questions, and seeks the answers, but an anthropologist cannot assume knowledge of what are the right questions. The anthropologist is not seeking a list of answers, but a list of *questions*. Getting the answers will be easy enough—once the right questions are known.

The anthropologist has a mixed bag of tricks for discovering the right questions. The fundamental trick is called "participant observation." As one anthropological handbook describes it, "Participant observation is field research in which the ethnographer is not merely a detached observer of the lives and activities of the people under study, but is also a participant in that round of activities." By living among the natives, pretty much as a native lives, the anthropologist hopes to understand the native's world view, after which the right questions will become more or less obvious.

Systems analysts also use the technique of participant observation, though not nearly enough. One of the reasons it's hard to define what a systems analyst does is that the systems analyst, like the anthropologist, must do whatever is required by the particular system being studied. An anthropologist studying the !Kung does different things than an anthropologist studying the Kippelers, because !Kung do different things than Kippelers do. Similarly, a systems analyst working on a credit clearing system does different things than a systems analyst working on warehouse automation, because credit offices do different things than warehouses.

A lot of the right questions become obvious when the analyst works for a few days handling credit applications or picking orders. Even so, there are limits to participant observation. For one thing, the analyst, like the anthropologist, is in danger of "going native"— identifying too closely with the system, thereby losing the perspective of the outsider. Analysts who work too long on the customer side lose their ability to see alternative ways of accomplishing functions, just as analysts who work too long on the data processing side lose their ability to see anything in noncomputer terms.

Another problem with participant observation is the amount of time it can take to achieve really deep understanding. Anthropologists know how to accelerate the process, and so should analysts. One of the most effective anthropological techniques that I've observed is the *meta-question*.

A meta-question is a question that directly or indirectly produces a question as an answer. For instance, you might ask a native,

"How would I ask someone for assistance in cutting hay?"

Or an analyst might ask a client,

"How would you get clearance on an applicant whose name appeared on the 'high risk list'?"

or

"How would you figure out what to do if just one package of an order didn't fit in the box you were packing?"

One of the most interesting meta-questions is one I used to use whenever I had to give an examination. It read as follows:

**Part 1.**   (50% credit) Write a question that would be an appropriate question for an examination in this course at this time.
**Part 2.**   (50% credit) Answer the question you wrote in Part 1.

This exam can save the professor a lot of work, because it can be used at any time, in any course, on any subject, and there's no possibility of cheating, even if the same exam is used several times in the same course. I use the same question in my analyst's bag of tricks, simply by asking the client,

"What do you think I should be asking you now?"

Again, I can use this meta-question many times throughout the interview, or in subsequent interviews. I've never met a client who didn't understand the question, or who got bored with being asked. It doesn't give all the information I'm going to need, but it's always a terrific way to start, or to proceed when we get stuck like two teenagers on a first date.

# Self-Validating Questions

Harlan Mills predicted that some day programmers will make so few errors that they'll remember every one they ever made in their entire career. I've had a long career, and I've made rather more than the one error per year that Harlan predicts. One a day might be more like it. But some errors were so gross or so costly that they stand out among the thousands.

Over twenty years ago, I was analyst/programmer for a service bureau studying a job that involved processing a million cards through the IBM 650 computer. Because of limitations on the 650's ability to read cards, the only punches allowed in the cards were alphabetics and numerics. Special characters could not be read at all.

When questioning the client in our very first meeting, I asked, "Are there any special characters in the cards?"

"No," he replied, "none whatsoever."

"Good," I said, "but I have to be sure. Are you certain that there are no special characters at all?"

"I'm quite certain. I know the data very well, and there are no special characters."

On that assurance, we went ahead with designing and programming the application, only to discover on our first production run that the system was hanging up on cards like this:

THREADED BOLT–1/2" #7

About sixty-five percent of the cards contained special characters, but when I confronted the client with this figure, he appeared genuinely puzzled. "But there are no special characters," he pleaded.

"Oh, no," I said triumphantly, "then what about this dash, slash, quote, and number sign?"

"Those? What's special about those? They're in almost all the cards."

Has twenty years of caution managed to repay the losses caused by my poor interviewing? I have learned to take a safer approach to framing questions. Perhaps if I share this approach with you, the score will be settled before I go before that Great Accounting System in the Sky.

In many books on interviewing, the reader is advised: "Don't use special language, or jargon that clients won't understand." That's lovely advice, but how can you follow it? The typical analyst isn't aware that there's anything special about the language being used—it's used in every interview.

Besides, it's not the language the client *doesn't* understand that kills you. It's what the client *does* understand, but in some other way. Like the word *special*. So we need a more powerful rule to avoid this kind of trouble. The rule I use is this:

### FRAME EVERY TECHNICAL QUESTION
### SO THE REPLY WILL BE SELF-VALIDATING

What do I mean by "self-validating"? The concept comes straight from information theory, and is analogous to the error-correcting and error-detecting codes we attach to stored computer data. Every reply is considered to have three parts, according to the formula:

REPLY = INFORMATION + VALIDATION + NOISE

For instance, suppose you ask the question, "How big is your inventory record?" and get the reply, "Oh, very long, about 600 bytes." You might analyze the reply as follows:

| Word | Meaning |
| --- | --- |
| Oh | Probably noise, but could indicate surprise at the question, which might indicate that the relevance of the question at this time is not understood. |
| very | A pure noise word, because you don't know what the comparison is relative to. |
| long | Mostly noise, but could say something about their experience at handling records of this length. |
| about | Validation of the *approximate* nature of the reply. |
| 600 | Although this seems purely informational, the two zeros suggest the approximate nature of the reply, as confirmed by the "about." |
| bytes | Information on the units in which the record is measured, but also validation that the question was understood in the terms it was asked. |

By contrast, consider the reply,

"625."

You can live without the noise, but without the validation, you are standing on quicksand. This curt reply *could* mean, as I once found out,

> "The record inventory we've ever had was 625 upright freezers that we had because we got caught by an unexpected model change."

This longer reply contains a lot of noise, but it's far superior to a plain "625," from the analyst's point of view. You may not like the information, but you know what the information is. Translated into usable terms, the validated information in the reply might be:

> "I don't understand the terminology of data processing, like the word 'record'. But I do know a lot about, and am concerned about, the inventory side of my business."

Analysts get into trouble when they try to eliminate the noise in replies by asking very precise questions. Very little of what passes for noise is pure noise to the attentive listener. The analyst shouldn't be afraid to frame questions that will generate noise in exchange for the opportunity to validate the reply. Replies to yes/no questions, of course, are very low in noise and thus in validation opportunities. That got me in trouble on the "special characters" problem, which might have been avoided if I had asked such noisy questions as:

1.  What characters can appear in your data cards?
2.  What are some typical cards, and some nontypical ones?
3.  How can I arrange to examine some of your actual data cards?
4.  Which of these characters (showing a list of all possible characters) appears in your data cards?

Notice the pattern of question words—what, how, which—instead of various forms of "is"—are, were, will there be—that prevent a one-bit, yes/no answer. With a little practice you can learn to frame questions in these more open forms, and to listen for the validation information among the noise.

But even yes/no replies can carry validation information, if we are attentive enough to pick up nonverbal clues. There are hundreds of ways to say yes, and thousands to say no. Hesitation time, facial expression, tone of voice—these are the most obvious ways to validate a simple yes or no. Don't try to make too much of these clues, because "body language" is pitifully crude as an information source.

But as a source of *validation*, these nonverbal clues can be used even by an unsophisticated analyst. When there is any sign of confusion, hesitation, or discomfort, use that information as a cue to frame a second question—one that will fetch up a bit more noise and a bit more reliability.

The hidden question you *really* want to ask is this:

"How reliable is your reply to the previous question?"

In some situations, you may be able to ask this question directly. If so, ask it. If not, try something else, like:

1.  What is the source of that information?
2.  Where can I find the documentation supporting that figure?
3.  What other information do you have that bears on the previous question?

In short, try not to be any less direct than necessary. Some people are offended by directness, but most are simply surprised by the refreshing change.

Personally, I'm more offended by indirectness. I'm especially put off by people who are trying to "read" my body language. I don't enjoy working with an interviewer who is trying to outguess me, especially when I'm supposed to be the client asking for a service. What I do enjoy is working with an interviewer who is skilled at framing direct questions and who shows interest in learning as much as possible about my needs and desires. And who isn't too bored by my "noise" to catch a few of my many mistakes. After all, even one error per year isn't perfect.

## "The Question Is . . ."

Systems analysts spend a lot of time exploring the behavior of others, but very little time analyzing their own behavior. In our consulting and training, we frequently put analyst/client meetings on video tape because it gives the participants a way of looking at themselves that's hard to achieve otherwise.

In reviewing these tapes, we begin to see patterns of non-productive or counterproductive behavior. A lot of this inept behavior centers around question-and-answer sessions. Here's a question about question-and-answer:

> "Do you think it's a good idea to ask questions that combine two pieces of information, or do you think that a question can have even more than two parts—unless it's really complicated—and would you rather stick with one form or the other, or should you use your own judgment except in those cases where some special rule applies, or when the question is so important that it demands a single expression, uncomplicated by other issues?"

Here are your possible answers:

1. Yes.
2. No.
3. Hmm.
4. What?
5. You jerk!
6. All of the above.

You may find the question ridiculous, but it's based on a transcription of an actual question asked in one of our sessions. The only thing different about the original was the presence of about thirty words of technical jargon that the user didn't understand, plus a few unflattering hints about the state of the user's intelligence.

How can such monsters creep into a simple interview? Like most monsters, they are born of fear—fear that the analyst won't do well in the interview.

Long, compound questions arise from fear that the original question wasn't sufficiently precise, or that the user didn't understand, or that there isn't time to ask everything that needs to be asked. They also arise from the analyst's fear of looking stupid,

71

which is why we sometimes shield ourselves with jargon or insults that project our stupidity onto the user.

Similar fears lead to transcripts like this:

"How often do you expect this report will be required?"
"Uh . . ."
"About once a week?"
"Um . . ."
"More than that?"
"Er . . ."
"Don't you even know how often you'll need your own report?"

The first error here is *not waiting for the answer.* Asking two questions without waiting in between is really no different from asking a compound question.

When we analyzed the tape of this meeting, we put the stopwatch to the intervals between questions. Between the first and second, there were 2.8 seconds—enough time to take a walk of five steps! And between the second and third, it was even shorter—1.9 seconds! What did the analyst have in mind by the rapid-fire questions?

He said he wasn't sure that the client had understood the original question. In that case, we suggested that he ask, "Is my question clear?" Instead, he managed to give the client the distinct impression that he thought she was too stupid to answer. And just in case that impression wasn't sufficiently distinct, he sharpened it with the wording of the fourth question: "Don't you even know how often you'll need your own report?" The bare question is impertinent, but the word *even* is the coup de grâce.

This analyst/client relation is turning sour faster than cider in a sauna, all because the analyst is afraid of the client. A slow answer could indicate stupidity, but if your client is that stupid, you have more problems than one question is going to solve. Why not assume that the delay indicates careful thought about a complex problem? If your assumption proves wrong, there will be time to correct it later—if you haven't hopelessly alienated your client.

Here are some other analyst practices we often find in "instant replays":

1. *Leading questions:* "You're not going to expand this file, are you?"
2. *Loaded questions:* "Are we required to preserve this clumsy format?"

3. *Self-answering questions:* "How many entries will there be in this table? I think fifty will be more than enough."
4. *Parting shots*—that last question asked after everyone has already stood up to leave, and that opens up new worlds of questions, all of which deserve sit-down answers.
5. *Controlling questions*—when the analyst breaks the client flow in order to return to some preset list.

   Deep down, each of these practices arises from the analyst's fear of not being able to *control* the interview. But why be afraid? The analyst and the client are working to the same purpose. The client is a human being—perhaps with the same fears of losing control. Interviews will go a lot better if such fears are brought out into the open, one human being to another.

   Consider the five questions above, and how they can be rendered harmless—even productive—by openly acknowledging your fears:

1. "I'm concerned about your future plans for this file, because if it expands beyond what we're now planning, we'll have to go to a two-volume setup, which will substantially increase the program's complexity. What can you tell me about it?"
2. "I had a hard time understanding this format, and I'm afraid that new users could have the same trouble. Also, it may prove costly to program. Can we consider changing it now, while we're redoing the system?"
3. "It's important to us to set a limit on the size of this table. The exact limit isn't too important for a table of this size, as long as we're sure it won't ever be exceeded in, say, five years. Do you think fifty entries would give us that protection, or should I allow more?"
4. "Something just occurred to me that didn't come up in the meeting. I know we don't have enough time now to deal with it thoroughly, so could I call you this afternoon when you have thirty minutes or so?"
5. "I'm worried that if we pursue that line of thought we won't get to some other vital things on my list. May I write it on the board as an unresolved area, to which we'll return at the end of this meeting or the beginning of the next?"

   As Humpty Dumpty instructed Alice, "The question is, 'Who is to be master?' " In trying to be absolute master of the interview, the analyst reveals his fear that he's not even master of his own house. Rather than trying to master interviewing by memorizing techniques and tricks, the analyst does best to master himself. The rest follows.

But how to master oneself? The Eastern philosophers have always understood, but it seems a most arduous lesson for us westerners. One masters oneself by giving up the attempt. By approaching an interview with the attitude that one cannot be absolutely in control, one attains the utmost possible control.

# Avoiding the "Plop Problem"

In the course of my travels last year, a computer user told me a story of an arrogant programmer. The user was a management consultant for a multinational manufacturer, and had been asked to evaluate an inventory management procedure that had been used with stunning success in their French operations. As part of his study, he wanted to compare the performance of the French procedure with other current procedures, using historical data from several countries. A programmer with a strong management science background was given the job of programming the simulation of the French procedure.

When the consultant received the results he could not reconcile them with the figures supplied by the French company. After extensive checking, he initiated a series of long phone calls to France, suggesting that perhaps their procedure had not actually performed as well as they had claimed. The French management soon took offense at this implication of incompetence and complained to the consultant's manager. Tempers mounted and international relations were strained to the breaking point.

By sheer chance someone examined the simulation program and noticed that the formula as translated into FORTRAN did not seem to match the formula as supplied by the French. One term was missing and a second term was negative instead of positive. The consultant, much relieved, took the program back to the programmer and showed him the error.

"That's not an error," the programmer protested. "Actually, the original formula was in error, so I corrected it. The formula I programmed is correct, whereas the original formula was simply wrong."

The consultant was astonished. "You don't understand," he said. "We're trying to simulate the precise formula used in France so that we can compare it to formulas used in this country. It's not a question of right or wrong, but merely of matching the existing formula."

"Well," the programmer replied, "anyone with a smattering of knowledge of inventory theory can see immediately that the French formula cannot possibly be correct, so what's the sense of programming it. Tell them to use *my* formula, if they want to improve their inventory management."

The management consultant decided to try another approach to the recalcitrant programmer. "That's a good idea. If you're right, I'm sure they'll really appreciate getting a better formula. In the

75

meantime, it will help them to accept the new formula if they can see how it compares with their original one on this data, so I'd appreciate having their formula programmed as soon as you can manage it."

"You don't seem to understand," the programmer insisted. "Why should I waste my time programming a formula I *know* is wrong? Just show them *my* formula and they'll understand."

At this point, the management consultant gave up on the programmer and got himself another one. The French formula was programmed and found to give the claimed results—which were, in fact, superior in many circumstances to the approaches used in other countries.

"And how did they compare with the original programmer's formula?" I asked the consultant as he finished his tale.

"Gee, I don't know. We never went back and looked at them. Do you think he might actually have been right?"

I replied that I had no opinion on the subject, as I had learned that arrogance and competence *can* occur independently in the same person, especially if that person is a programmer. Many programmers have a double standard in these matters:

1. If a user makes a mistake, the user is stupid.
2. If a programmer makes a mistake, the job is difficult (probably because the user is stupid).

In other words, the same programmer who is all too eager to take responsibility for the user's part of the job cannot be pinned down when it comes to placing responsibility for the programmer's part.

I frequently hear the complaint from nonprogrammers that "programmers are irresponsible—they don't take responsibility for anything they do." I believe this complaint to be a half-truth—quite dangerous to a productive picture of the programming task. Although some programmers shirk responsibility, others commit much worse errors by assuming responsibility where they have no justification, as in the consultant's story.

It may be edifying to ask how this arrogant programmer got that way. Was it a defective gene, or was it something learned?

Lacking sufficient background in genetics, I'm inclined to the learning theory of arrogance. I can imagine that this programmer has had a bad experience in his past which led him to adopt the arrogant posture as a defense. What sort of experience could that be?

Most programmers can readily answer that question, for the experience is universal and frequent. Given an assignment that seems a bit "off," the programmer attempts to question the analyst or user. With little ceremony, he is told to stick to what he understands—to follow the specifications as given. Much later, when he delivers the requested program, the analyst and user are furious at the programmer for doing what they *said*, rather than what they *meant*.

And should the programmer protest by saying, "I told you so," they'll reply by saying, "If you knew it was wrong, why didn't you insist that we listen to you?"

If you've been a programmer for more than a week, you probably winced at that last line—I always do. And if you've lived through that cycle more than three times, you probably defend yourself by using the approach of the miserable programmer in the Case of the French Inventory. Looked at from this perspective, that arrogant wretch seems rather humble, reasonable and professional. All he was doing was trying to *avoid a problem before it became a problem*—a magnificent idea that is almost always preferable to the "problem solving" we hear so highly praised nowadays.

The problem the programmer was trying to avoid is one of the great classic problems of programming. We call it the Plop! problem, after an image from Chaplin's film masterpiece *Modern Times*. Chaplin is in the penitentiary (although utterly innocent, naturally), seated with hundreds of identical gray prisoners at the long mess table. Down the table, moving rapidly in Chaplin's direction, is an inmate dishing out huge ladles of an amorphous, gray, glutinous mass that would make haggis look like smoked salmon. Just then, Chaplin's fork falls on the floor, and while he is bending over to retrieve it, a gray glob of this delicacy is plopped onto Chaplin's plate. Chaplin finds the fork and sits up, suddenly noticing the substance that has materialized on his plate. He looks at the mass, puzzled for an instant as to its source. Then, in one of the most magnificent moments of the silent screen, his face lights up and he raises his eyes toward the heavens, in hopes of spotting the flying beast who has deposited this memento, Plop!, on his expectant plate.

I recall this scene every time I see some programmer blindly, faithfully, and silently accepting some assignment that has seemingly plopped on his desk from the heavens. It's as if they believe that by "eating" this assignment without protest, they avoid all responsibility for any indigestion it ultimately produces. When it comes time to distribute the blame, they can truthfully claim, "I was just following orders."

In my opinion, the truly professional programmer must tread a narrow path between accepting everything that happens to plop on his desk and rejecting everything that has the slightest error or confusion. The French Inventory programmer refused to be plopped on, thus avoiding one problem, but refused in a manner that created another. Indeed, I'd like to suggest that this kind of arrogant front is actually only a more sophisticated approach to avoiding responsibility. By acting in a sufficiently obnoxious manner, you can ensure that nobody will listen to your objections, regardless of their merit. Then, if things come out wrong, you can remind everyone that "I told you so," while if things come out right after all, you can count on nobody remembering what it was you (incorrectly) objected to.

Is there a better way? Or better, are there better *ways?* I believe there are, for many good programmers seem to have found a way to steer a middle course. Here are some suggestions gleaned from discussions of the problem in our technical leadership workshops:

1.  *Watch your language.* What seems like arrogance can often be poor choice of words. Instead of saying, "This specification is wrong," try saying, "I don't understand this part of the specification." Nobody can say, "Yes, you do understand it," and no reasonable person can take offense at this approach.

2.  *Broaden your understanding.* Unfortunately, specifications rarely give reasons for the actions they specify, and many curious parts of a specification can be understood only in terms of the bigger picture. Spend a little time getting to know your user in general terms, and where the application fits into the scheme of things. Go out into the world of the user to see how things actually look, rather than how they are translated through the specification. Reasonable users will appreciate your interest, and be proud to show you what they're doing.

3.  *Be tolerant of imperfection.* Any programmer knows that everybody makes mistakes. You don't want to accept mistakes, but you should be understanding of the people who make them. Even if you, yourself, are perfect, you'll frighten other people if you tread heavily on them each time they err in their dealings with you. Soon they'll either avoid you, or seek to catch and discredit you as soon as you make a mistake.

And even if you're never caught, the time spent in playing these games is wasted for more fruitful pastimes.

4. *Accept the final authority of the user.* In the end, your job as programmer is to provide a service to someone else. After exhausting all reasonable avenues, you may still have a difference of opinion about the "right" way to do something. Your user pays the bill, and is therefore the final authority— on this job. Don't do things that violate your moral code, but no reasonable user will ask such things of you. And if you are repeatedly "smarter" than your user, perhaps you should be in his business.

5. *Don't work for unreasonable people.* If, after repeated tests of the situation, you decide that your user or analyst really is pigheaded, slip out of the assignment, get transferred, or get another job. Pigheaded people will rarely give credit when you're done even when things work out right, so why waste time in a lost cause? But be gracious about it, for when other people seem pigheaded it usually turns out to be a matter of incompatible personalities, which is just a fact of life we sometimes must accept.

By applying these problem-avoiding techniques, our students have managed to avoid getting plopped on, yet still produce a great many satisfied users.

# Avoiding Communication Problems through Generalization

To many people, programmers are poor communicators. In many cases, the programmer fails to understand what a user *means,* and merely does what that user *says.* Some say that programmers resemble, in this way, the computers they serve, thus demonstrating that the programmer's true master is the computer, not the user.

I personally feel that this view of the programmer's communication skills is obsolete, and also oversimplified. The modern programmer is more likely to err by *guessing* what a user means, rather than following literally what is said. If you are a programmer, you can test my theory by asking yourself how you would interpret the following statements if presented to you by a typical "naive user":

1. There can *never* be any codes besides 'A', 'B', and 'C'.
2. The largest table of words that could possibly occur has 1,000 word entries.
3. I don't know how many different part numbers there are.

The experienced, modern programmer responds to these three specifications by thinking:

1. We'll put the codes in an expandable table, and try not to presume that the new codes will always be a single character. We can probably get away with a linear search of the table.
2. He's guessing wildly about the future size of this table, so I'll make it dynamically adjustable. If I'm terribly pressed for space in the initial system, I can probably get away with a few hundred words, as he's unlikely to reach the 1,000 in the near future, if ever.
3. The addition of new part numbers is an uncontrolled process, which means that I'll have to allow for a very large file, both in space and in processing methods. But more important, the form of the part numbers themselves is unlikely to be confined to the forms that exist today, so the program had better be able to accept almost any variant form of part number.

These are but a few of the thoughts a modern programmer has in response to such statements, but they may be meaningfully contrasted with the thoughts these same statements might have generated ten, fifteen, or twenty years ago, when machine capacities were much smaller and machines were much slower:

**1.** I can code this test as

> IF CODE = 'A' THEN GO TO PROCESS-A.
> IF CODE = 'B' THEN GO TO PROCESS-B.
> PROCESS-C. . . .

Furthermore, I might save some time by reordering the tests to put the most frequent code first.

Notice how logical this translation is when machine speed is at a premium, along with space. Notice too how the user will first discover that when a code of 'D' happens to come along, it is handled as if it were a 'C', which isn't what he meant at all. Then, to make matters worse, the user finally adjusts to 'D' meaning 'C'—because he's been told that "there's no way the program can be modified without causing more trouble." Then the programmer changes the test to put 'C' first, in order to "make the program faster" in response to rising volumes. Suddenly, 'D' begins to mean 'A', or 'B', and when the user asks why, he's told that "we had to modify the program." If he's foolish enough to say, "but I thought you said the program couldn't be modified," he will earn the everlasting contempt of the programmer for not understanding the difference between a "frill"—like handling code 'D' separately—and an "essential"—like cutting thirty seconds off the CPU time.

Or consider how the second example would have been interpreted in those dear dead days of long ago:

**2.** There's no way I can ever spare 1,000 times $N$ characters, when core is already getting close to the edge. I'll talk him down to the 200 or so he really needs right now, and convince him that $N = 8$ is sufficiently long for any reasonable word. Then if I can sort the words by length, I may be able to save even some of those 1,600 characters. Or perhaps I can sort by first letter, and store only the last seven characters for each word. If I spend some time on it, I'll bet I can get this table down to under 1,000 characters, which should be okay for the time being.

Again, on a machine with 10,000 characters of memory for data and instructions, the programmer is quite likely justified in making this interpretation. If the user wants his application to run at all, he'll have to accept the conditions placed on his word list. He should be able to understand that space is at a premium, so that words cannot be added willy-nilly to his list.

Thinking he understands about space, he may imagine that he could at least substitute a new word for an old one, but is dismayed to learn that "because of the complex data structure," it would require "practically reprogramming the entire job" to make such a change. If he persists in asking for an explanation, he'll get an earful of "threaded lists" and "the superior efficiency of hard-coded constants"—adding further support to the image of the uncommunicative programmer.

The third example is no better, for when the user says frankly "I don't know how many different part numbers there are," the programmer is free to think:

3.  If he knows so little about his application, he's going to have to live with some restrictions. First I'll have to put the file on tape, since expanding a direct access file will be murder. Second, the format of these part numbers will absolutely have to be fixed, because we can't be rewriting all the old tapes just because somebody wants to try something new. Besides, we don't have a sort for variable-length keys—unless we're willing to put up with intolerable inefficiency, and a few lost records from time to time.

The user is told that, under the circumstances, he can't use his new on-line terminal for accessing information about parts. He'll have to wait until the end-of-week update run, by which time the information will be useless. Under the circumstances, he'll have to give up all hope of controlling the assignment of new part numbers—one of the major reasons he got this computer in the first place. When he complains, the programmer wails, "But how can you expect us to put this information on-line when you don't have any control over the assignment of the part numbers?"

Contrast the old and new interpretations of these questions and you'll see that the programmers haven't changed their communication pattern one bit. The problem they face is a difficult technical environment that they couldn't possibly explain to their user. To avoid problems, they interpret the user's specifications in that way which will cause the least trouble, either now or in the future. They don't get any credit for the problems they avoid, but only blame for the lesser problems that might be created as side effects to their best efforts. And then the unappreciative users call them "uncommunicative," or something worse.

Yet if we examine the "old" and "new" approaches, we begin to see a promising trend. In the old days, the motto was:

ANY AMOUNT OF PROGRAMMING TO SAVE
ANY AMOUNT OF MACHINE

Gradually, under the strong influence of advancing technology and the subtle influence of the hardware salespeople, the motto became:

ANY AMOUNT OF MACHINE TO SAVE ANY
AMOUNT OF PROGRAMMING

But even that reversal didn't go far enough. Little by little another change took place. In the old days, another motto was:

ANY AMOUNT OF TROUBLE LATER TO SAVE
ANY AMOUNT OF TROUBLE NOW

Then, under the influence of Father Time, "now" became "later" and the price had to be paid. With the wisdom of hindsight, the motto became:

ANY AMOUNT OF TROUBLE NOW TO SAVE
ANY AMOUNT OF TROUBLE LATER

But the trouble was not just with the machine. More often than not, it was with the user—who knew the programmer was "uncommunicative" and therefore not worth trying to communicate with. Little by little, programmers learned the wisdom of another motto:

WHEN YOU CAN'T FIGURE OUT WHAT THE
CUSTOMER WANTS, GIVE THE CUSTOMER
EVERYTHING

This tactic, often called "generalization," avoids communication problems by substituting both programming and hardware.

But sometimes the price is too high, given the state of the art in hardware and software. Sometimes, when we provide everything, we wind up with nothing. The package may prove too expensive for large classes of users. Alternatively, there is so much potential function that nobody can understand how to do simple things with the program. So the tactic of generalization cannot be applied blindly. For instance, if we allocate an array for all possible parts and allow full flexibility in code size, word length, and part number format, we could well exceed the capacity of even the most capacious virtual memories. I'm reminded of a clever programmer many years ago who told his user—a manufacturer of complex equipment—that parts explosion "was nothing more than a big matrix multiplication." His flippancy encouraged the manufacturer to believe that his application could be programmed easily and run efficiently. The programmer had neglected to observe that since there were over 10,000 different parts in use, the "matrix multiplication" would involve a matrix with 100,000,000 elements.

Nevertheless, with a bit of modesty, caution, and humility, generalization can be a powerful substitute for communication. The analyst/designer, in making a list of questions for the user, ought to consider for each question:

1.  Can the user actually know the answer to this one?
2.  If the user can know, will it cost more to find out than to design in such a way that I don't have to know?
3.  If I can design so that I don't have to know, is it possible that this design is actually cheaper than a design that would depend on knowing?

If the answer to any of these three questions is yes, then it should be better to use design—generalization—rather than communication—questioning the user.

It may be surprising to some people, but problem solvers in other fields have long known, as George Polya said,

"THE MORE GENERAL PROBLEM
IS OFTEN EASIER TO SOLVE."

In the case of a program, the more general design may be easier to implement, may operate faster on the machine, and can actually be better understood by the user. In addition, it will almost certainly be easier to maintain, for often we won't have to change anything at all to meet changing circumstances that fall within its range of generality.

Unfortunately, many of today's most widely used programming tools tend to discourage generalization, by making it clumsy or inefficient. Perhaps by elevating generalization from a "dodge" to a "strategy," we would encourage our tool builders to give it more support.

## The Fairy and the Pig: A Fable

In Heaven, there is a schedule by which each fairy gets to go to Earth from time to time and grant a wish. Though the highest ranking fairies may make the trip as often as once a month, some are so cynical from previous wish-grantings that they go only once a year, usually in the fall to see the new fashions. The younger fairies, however, have more enthusiasm and never miss an opportunity to make life on earth a little closer to their own heavenly existence.

The youngest fairies can only make the trip on the twenty-first of June, and this year their excitement built up something fierce as the end of spring drew near. Plans for the impending trip so dominated their conversations that older fairies betook themselves to cloud regions far from Heaven's exit port. But the young ones didn't notice, so busy were they exchanging tales and plans. Had this not been Heaven, an observer might have thought that each was trying to outdo the next.

There was considerable competition in trying to make an original choice of subject. Should they bestow their wish on someone poor or someone rich but unhappy? Should the lucky one be ugly or pretty, grouchy or merry? In the midst of these debates, one of the very youngest fairies ventured to ask, "Why do you always grant your wishes to human beings? Don't animals have wishes too? Personally, I'd like to grant a wish to a pig."

The other fairies didn't take these remarks seriously, so the subject was dropped from the conversation. But our young fairy didn't forget so easily, and when midsummer came, he betook himself to a farm near Pocahontas, Iowa, there to alight on the wooden railing of an enormous hog wallow. The pigs, as is their nature, regarded him curiously until satisfied that he couldn't be eaten, at which point they gave a collective grunt and returned to their wallowing. Our fairy, to say the least, was overwhelmed with disappointment.

But one of the pigs didn't behave like the others, and waddled over to the fence. Our fairy took heart and addressed him thus, "Mister Pig, I see that you are a hog apart. Would you like to know who I am and what is my mission?"

"Sure," snuffled the pig, trying to smell the answer with his nose, which he trusted more than words anyway.

"I'm a fairy, and I've come down from heaven to grant someone a wish."

"A fairy," thought the pig, who was a cut above the others when it came to brains. "No wonder I didn't recognize the smell."

"Do you know what a wish is?"

"You bet your sweet life I do," the pig snorted. "Every minute of every day I wish I had a pair of wings like the birds—and like you—so I could fly out of here and not have to share this wallow with those stupid swine."

"How perfectly marvelous," thought the fairy. He raised his twinkling wand and passed it three times over the pig's back, saying, "Then you shall have your wish, your heart's desire." And lo and behold, a pair of wings began to grow between the pig's shoulders.

"Wow," he grunted, "you really did it. Can I fly with 'em?"

"Try them and see," the fairy replied. "The horizons of the world are now yours for a mere flutter."

The pig shook his wings, awkwardly at first, then gaining better control, he rose above the pigpen, soaring past the fairy who had to strain to catch up.

"Do you like them?"

"Sure do!"

"And where are you going first?"

"Right here," the pig answered, splashing down in a wallow about a stone's throw from the one he had just left.

"Here?" the fairy asked. "Why here?"

"Cause Farmer Brown never lets us go here unless he forgets to close the gate. This way I don't have to share this wallow with those stupid swine. I have it all to myself." And with that, he rolled over in mud so deep his beautiful new wings became as sloppy as the rest of him, and totally incapable of flight.

MORAL:    *If pigs had wings, they'd still be pigs.*

IN OTHER WORDS: Interviewing is always limited by your customer's knowledge, just as a systems design is always limited by your customer's values.

# PART
# V

# Design Philosophy

*The Goat and the Hippo: A Fable (page 106)*

## A Simple View of Design

*It does not matter whether one paints a picture, writes a poem, or carves a statue, simplicity is the mark of the master-hand.*

Elsie De Wolfe

What is *design*? The answer isn't simple, but there are many simple answers. One of the most useful is a model of design based on the process called *natural selection*. Here's how it works, in words and pictures:

A system exists in many copies, each one somewhat of a variant about some central "type":

Each copy, being slightly different from each of the others, reacts differently to the common environment:

Copies that react more favorably produce other copies, more or less like themselves:

89

But always with slight variations from the original:

Less successful copies produce other copies, too, but not quite so many:

All this is spiced up with a generous pinch of randomness, so that from time to time a rather poor copy happens to make out well:

and a rather good copy gets clobbered:

In the end, though, there are so many copies that the Law of Large Numbers begins to work against those variations that, on the average, aren't too hot:

There is, of course, much more to the theory of natural selection than this bare outline, and there are also many other change processes going on in the world. But these other change processes do not meet one of the essential conditions required for natural selection to take place:

1.  There must be replication with variation.
2.  There must be selective action of an environment.

In what ways, then, does the process of design parallel natural selection? We can fit the process of design to the model of natural selection by breaking the design process into two different kinds of step:

1.  Generation of new ideas
2.  Selection among these ideas according to criteria.

The ideas correspond to the variations and the criteria to the selective environment. Although we may actually be building only one system, we may produce hundreds or thousands of conceptual variations before we choose the set we shall actually implement.

In biological evolution, of course, the testing of each variation is actually carried out in nature, as a life and death process. Viewed in terms of cost in living material, it is a slow and expensive process. Design, on the other hand, is *not* carried out—insofar as we can keep it from being carried out—on the actual system. Instead, we create images, or perhaps models on paper, which are far cheaper to build—and thus cheaper to kill if they are unsuitable.

For the process of design to work, we must be able to evaluate these images or models without actually building the systems they represent. *Learning to design is learning to generate and evaluate models.* We have neither the time nor the energy to design systems through the process of natural selection on actual systems. In extreme cases—where we have no criteria on which to evaluate competing models—we may actually implement two or more of our ideas, but this would be a last resort.

This natural selection analogy makes us aware of the twofold and paradoxical nature of design. We must have ways of generating ideas (which we might call *synthesis*) and ways of getting rid of ideas (which we might call *analysis*). When a designer omits one or the other of these twins, the result is a characteristic misdesign. Perhaps one example of each will help.

Synthesis without appropriate analysis is perhaps most conspicuous, for we see the wild idea become a monster whose behavior has not been predicted with sufficient care. A data base system we recently reviewed was found to have a capacity of only 5,000 transactions per shift—instead of the 50,000 that had to be processed. The designers had failed to analyze the consequences of their design, which had piled interface upon interface for programming convenience, but accumulated a slowdown of ten times in the process.

Analysis without appropriate synthesis is harder to detect, for we rarely see the good ideas that we never thought of. In another data base application, the designers analyzed their approach and found that a conventional disk drive would not have sufficient capacity. This analysis in hand, they proceeded to order a specially built second access arm for their drives—a custom-built system that cost a lot and gave lots of trouble operationally. Stimulated by the equipment failures, another designer conceived of a different way of organizing the data base—a way that made the second arm entirely redundant. The expensive hardware was removed, and the system settled down to what it might have been had the designers troubled themselves to generate a few more possibilities before ordering special hardware. This then, is the essential injunction of design: *Don't try out your ideas without sufficiently analyzing them, but don't waste your time analyzing when you don't have a good enough idea to begin with.* Put into a short and sweet motto that alludes to its biological origins, this says:

STRIKE A BALANCE BETWEEN VARIATION
AND SELECTION.

And how do you strike this balance, in practice? Simply ask yourself, "Do we have too many ideas, or not enough?" If you have too many, move into analysis to reduce the excess. When you finally have too few, move back into synthesis, to generate more. Keep oscillating until the process converges—then you have your design. It's that simple!

# Thing versus Process—The Grand Dichotomy

*There are two kinds of people in the world: those who divide every-*
*thing into two kinds, and those who don't.*

Kenneth Boulding

There's no use attempting to conceal the irresistible human temptation to dichotomize. Rather than hide our vices, we ought to
study them. When we have a strong genetic urge to do something—
like get together with members of the opposite sex—it's because
that urge has served some useful function for the species in the
past. Today it may be counterproductive, but it had to have some
pressing value in other times and other places.

I believe the vice of dichotomizing evolves from the great
power of the "divide and conquer" strategy in a world dominated
by the Square Law of Computation. The Square Law says, in one
context, that if we try to go from A to B, the complexity of the
change goes up as the square of the "distance" between A and B.
The more our future system, B, differs from our present one, A,
the harder is the design and development.

The Square Law is constantly thwarting our attempts to design a better system. Our basic tool for defeating it is to divide the
trip from A to B into two trips, from A to X and from X to B.
Instead of designing a single system (B) to process living data and
produce management reports, we first produce a system (X) that
filters the living data and produces purified data for another system
that produces the reports. By dividing the problem in half, we have
to think about only one half at a time—which is four times easier
than thinking about the whole. Changing a system one step at a
time simplifies the designer's problem and lessens the burden on
all people involved with the system.

But not every desirable design can be achieved through a
series of small changes to an existing system. When crossing the
Grand Canyon on a motorcycle, two shorter jumps won't work. In
design we will always have to make conceptual leaps from what
exists to what might exist. There will always be the risk that one
of the leaps will fall short, and plummet us into the abyss.

The first level of incremental design is a series of *real* changes
to *actual* systems. The second level of incremental design is a series
of *imaginary* changes to a successful *general design scheme*. Designers minimize their risks by passing first to the second level of
incremental design—mental changes to design schemes. Each designer carries a mental bag of such conceptual design schemes, each

subject to constant variation and refinement. Through variation and selection, the population of schemes in each designer's mind evolves. Through many small experiences, patterns of thought crystallize into design techniques—tactics for winning small design battles.

But the design war isn't won through winning all the small battles. After many battles, larger patterns emerge—grand strategies of design. It is not surprising, then, that each of these grand strategies is related to some general mode of human thought. For instance, one of the greatest divisions is between static and dynamic schemes, between thing and process, between noun and verb. Humans, being animals capable of and dependent on locomotion, are greatly attracted to *process* as a way of coming to mental grips with a subject. We distinguish the dog from the cat not on the basis of anatomical differences—which are not lacking—but on the basis of a few *behavioral* differences. Beverly purrs, stalks mice, and seeks human company only with diffidence. Rose barks, tears after rabbits, and jumps up and kisses strangers. Which is the dog and which is the cat? There can be no error, yet it would be hard to tell them apart with such a brief anatomical description. From a distance, we might perceive Beverly as a small dog—until she takes her first controlled step, or stretches her spine.

It is the same with the description of computer systems. An APL and a BASIC user's manual look very much alike—at least the task of rummaging through them for information is equally unappealing. Yet a few minutes interacting with either system will establish behavioral differences between them as unmistakable as a purr and a bark.

To designers, process description schemes have this same lively appeal. One of the easiest ways to get involved in the design problems of an unfamiliar system is to run through imaginary scripts, or—to use a favorite term of American politicians—scenarios:

"The user will do so and so."

"Then the system responds with thus and such; and the user comes back with . . ."

It is so natural to do this that we may not think of it as a method at all, though every designer employs it as a major tool.

By itself, the system's behavior is certainly not a reliable basis for design. For reliability, we must have a static *structure* that we can study at leisure and come to understand as a complete entity, not as a disjoint set of transitory behaviors. This, indeed, is one of the central insights of the structured programming movement—that we must break away from dependence on the intuitively appealing but incomplete process description. We must design structures that can be shown to engender the complete process we desire.

Yet the very incompleteness of the process description is what makes it so effective in the early stages of design. Any system contains far too much information for us to digest all at one sitting. We don't form an initial impression of cats by reading a textbook on comparative anatomy; we pass an idle moment glancing at Beverly, Henry, and Vivian doing mock battle among the footprints in the snow.

Similarly, we don't approach APL by reading the system description manual; we pass an idle moment toying with the terminal. For those who want to understand APL more fully, there is a shelf of manuals standing and waiting to serve. (I don't have any Beverly manuals, but who will ever understand a cat by any means?)

Each of us tends to have a preferred mode of obtaining information about a system, and of describing a system. But the lesson from nature is an important one:

NEITHER A STATIC NOR DYNAMIC VIEW
CAN BE THE WHOLE VIEW

Please keep this lesson in mind the next time someone tries to sell you the one all-purpose systems design/analysis approach. If it doesn't embody both *thing* and *process*, get yourself a cat, instead.

# The Three Bs

*Of systems, the minutest crumb*
*Must Be, Behave, and then Become.*
*This Principle the space traverses*
*From Atoms up to Universes.*
*And systems that are not malarky*
*Must find their place in this hierarchy.*

Kenneth Boulding, commenting
on an essay by Ralph Gerard

Although the grand dichotomy is between thing and process, the designing mind needs more aid than a single division can provide. In Gerard's words, each system has:

1. Constant architecture in time, its morphology or "being"
2. Reversible changes in time, its functioning or "behaving"
3. Irreversible changes in time, its development or evolution, or history, or learning, or "becoming."
   (See Figure 3).

Quite often, designers consider only the middle member of this set—the functioning or "behaving" of the system. First, they ignore "being." But a system requires a lot of attention to stay constant in a changing world. Indeed, for many natural systems, maintenance is the primary activity. Designers also ignore "becoming," yet the act of making a new system—or of changing an old

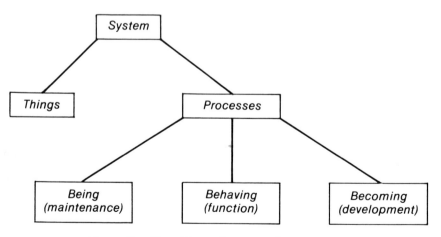

**Figure 3.** Two Great Conceptual Divisions

96

one—is also a behavior. It is the behavior of another system that we call *the system development process*. Taking this view, system design can be approached at two levels:

1. Designing a system with a particular behavior
2. Designing a system development process that will itself build the system with the desired behavior.
   (See Figure 4).

There is much interest these days in automatic systems for system development. With such a development system, you need only describe the desired system, feed the description into some computer, and wait at the output for your neatly wrapped package. The prospect sounds so wonderful to harried DP managers that any programmer who claims to know how to write such an automatic system is in danger of drowning in cash.

There's a certain irony in paying programmers to get rid of programmers by writing even more complex programs. It may prove to be progress, but it reminds me of how nineteenth century doctors managed to "solve" the rather innocuous social problem of morphine addiction by introducing heroin! But even if automatic aids to system development are a symptom of addiction to technology, they aren't necessarily harmful. I confess to the same addiction, and I believe I wouldn't be one-tenth as productive technically without many development aids I've created or borrowed. But I worry, all the same, about this tendency to dream of complete automation of software development.

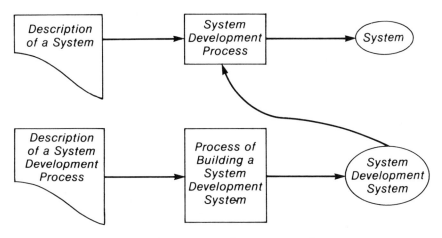

**Figure 4.** System Development as a System

Strive as we may for automation, programmers are sure to remain for a few more years. At least in the short run, the successful designer must realize that describing a desired system is only a third of the battle. The day may eventually arrive when such a description can be fed to an automatic system builder, but for the present, the designer must be concerned with numerous details of how the system will actually be assembled and maintained.

We've seen this situation before, as in the history of architecture. When a Frank Lloyd Wright decides to build a spiral building, like the Guggenheim Museum in New York, he tackles the job on all three levels—being, behaving, and becoming. If he were building another rectangular box of conventional design, he could count on the existing "infrastructure"—a vast army of workers would transform his descriptions into reliable buildings, which other workers can keep operating.

But he wasn't designing a conventional box, so Wright had to do much more. None of the infrastructure existed for a spiral museum, so although he began with a description—far more detailed than usual and expressed in nonstandard terms—Wright wasn't finished there. During construction, he had to pay frequent visits to the site, and even after construction he had to linger to solve such mundane problems as the traffic pattern caused by the spiral and the leaking roof caused by new techniques and materials.

Not all of our information systems need be spiral museums. The model that has tempted many of those who would automate development is the "mobile home." It is built in a factory, using standard factory methods, and can presumably be maintained with a screwdriver. Because this infrastructure exists, the mobile home designer need only draw standard blueprints and pass them to the factory. After a short delay, mobile homes are ejected on the shipping dock, ready for standard transport to a standard foundation in a standard mobile home park.

Until now, most of our information systems have been at the level of spiral museums, rather than rectangular mobile homes. No doubt there will be some lowering of "taste" when all the costs of custom design and implementation are reckoned. No doubt application packages will form the mobile home parks of the future, but people who can afford "old-fashioned" houses aren't flocking to live in mobile home parks. Even after a quarter century of mobile home development, about all we can say for this style of living is that it is cheap. Why? Because that cheapness was obtained only partly through automation of the system development process. More of the cheapness comes from ignoring the other two processes:

1. Behaving—that is, giving the homeowners what they really want instead of what can be produced by factory methods
2. Being—that is, paying for maintenance, instead of allowing the mobile home park to become a slum in three years.

Of course, in many cases we've been paying for Frank Lloyd Wright museums but getting slums—with one or more of the "three Bs" left out of the equation. In such an environment, the prices of mobile homes start looking more attractive, so unless designers get "Wright," they may find themselves living in slums themselves.

For the foreseeable future, the system designer who wants to stay employed will have to wrestle with all three aspects of "system as process." If you want to be a great composer, study your Bach, Beethoven, and Brahms, but if you want to be a great systems designer, concentrate on the other three Bs. At least for systems that are not malarkey.

# Design for Understanding

Here is the opening of Chekhov's story, *Ionych*:

> When people who had lived only a short time in the town of S___
> complained of boredom and the monotony of life there, the local
> inhabitants, as though in self-justification, claimed that on the con-
> trary there was nothing wrong with S___, that it had a public library,
> a theatre, a club, that balls were given there, and, last but not least,
> that there were quite a number of intelligent, interesting, and pleasant
> families with whom one would strike up an acquaintance. And they
> would point to the Turkin family as one of the most cultivated and
> talented.

What an incredible picture of the provincial town and its life the
master paints with the words of a single sentence! How much he
loads onto the canvas with a single careless stroke such as "a club"!
And how well, with a simple second sentence that crashes into our
simple brains, he describes the Turkin family and places it precisely
at the right spot in the composition! We hacks can never aspire to
the genius of a Chekhov—not to speak of a genius that even survives
translation to another language. We can, however, learn from studying his work.

But the true mastery of a Chekhov is that so few of his
readers study his work at all. To them the work itself is transparent.
They are not reading words about a country town, they are *in* a
country town, experiencing it. They read, they enjoy, they learn
many interesting and useful things, but if they perceive the hand
of the writer in the work, he has failed. And so it is for the system
designer. Our best work passes unnoticed, while our failures attract
attention.

We do not know if Chekhov was actually in the town of
S___ when he wrote, or made it up from memory, or designed it
from elements of other towns. Whatever method he used, he managed to capture our common memory of whatever small towns
we've known, even towns half a world away. He has reached the
common level of ordinary experience, and so is a great writer, loved
everywhere, even in translation. Chekhov, we say, has "the common touch."

Systems designers, to be successful, if not loved, must also
have the common touch. Good systems, for example, because they
must be maintained by ordinary programmers, are far less sophisticated than the designer is capable of producing. The designer who
has contempt for the "common programmer" will fail, for contempt

will leak out in the form of irresistible displays of unmaintainable erudition. Instead the designer must work in a deliberate, conscious, and controlled way to ensure that the level of the product is never inconsistent with its survival in a real environment of real people with real limitations.

This kind of design is not easy. Some design theorists have implied that the only design problem worthy of intellectual consideration is the design of ever purer algorithms. Nobody can disparage this 'computer science' activity, for today's science becomes tomorrow's engineering, and tomorrow's engineering becomes the day after tomorrow's common fieldhand knowledge. But designers are not designing for the day after tomorrow. Most of the time, they are lucky not to be designing for yesterday.

Besides, *common* does not mean *simple*. It's far more difficult to write a readable story on a small-town family than an erudite work on mathematical esoterica. Similarly, the truly challenging job in system design is to

## DESIGN FOR UNDERSTANDING

In order to design for understanding, the designer must possess an even deeper level of understanding, just as the great writer never reveals the incredibly complex thought and labor underlying the simplest opening paragraph.

And how does one learn to design for understanding? How does one learn to write with the common touch? Nobody knows, but we do know that all great writers loved words, and worked with words ceaselessly. Reading Chekhov, even without conscious effort, we become better writers, and better readers. Studying well-designed programs, we become better programmers, and after years of study, we become more capable maintainers.

Slowly, but inevitably, the level of program design can be raised as the level of the average maintenance programmer is pulled up. The "common touch" is common only to one period of time. Three centuries ago, there wasn't an educated reading audience large enough to support great writers. The writers and their readers grew together, and created a common level of communication. So too did agricultural engineering grow hand in hand with the education of the average farmhand. And so must system design grow hand in hand with programming, and with the population of system users.

The architect without the stonemason is not designing cathedrals, but castles in the air. The program designer without an appropriately skilled programmer is doing even worse. Nobody would dream of trying to build a cathedral without skilled masons, but programmed systems are treacherously "invisible." Anyone can see that the apprentice mason cannot easily produce a smooth face or a straight edge, but to the unpracticed eye, one program listing looks pretty much like another. Thus it will be that a horde of "warm bodies" will be assembled to produce "the system," or, even worse, to maintain it after the master programmer has finished "building" it.

We've all heard of the importance of "self-documenting" programs, but designs too must document themselves. They must not achieve this documentation by slapping hundreds of crude words onto an equally crude design. The best paragraph does not seem to be written, and the best design does not appear to be "documented." Instead, the design itself embodies its own understanding—to look upon it is to know what it can do, what it cannot do, and what we can do without destroying its essential nature.

A self-documenting design will not encourage someone with inadequate skills to attempt repairs. On the other hand, a self-documenting design will draw the attention of the skilled programmer toward some untidy part that needs work. The true test of a self-documenting design is time. It must encourage each hand to maintain not just its form, but its spirit. It will not only be maintained, but *its maintainability will be maintained.*

If a program is to be maintained by sophisticated chief programmers in Gotham City, then it may be designed in a sophisticated way. But for your "intelligent, interesting, and pleasant" programmer in S____, perhaps a different design is more appropriate. Only a genius can write one design for both—a design that neither will notice.

# On the Origins of Designer Intuition

*I learned three important things in college—to use a library, to memorize quickly and visually, to drop asleep at any time given a horizontal surface and fifteen minutes. What I could not learn was to think creatively on schedule.*

Agnes De Mille

Designers used to be born, or perhaps self-made, but now they are being manufactured assembly-line fashion by colleges. Yet colleges still don't know how to teach people to "think creatively on schedule," so college-trained designers face a dilemma. Designers, every day, have to generate new creative ideas, but our limited brains make it virtually impossible to imagine anything *completely* new. Or, if we succeed, to figure out how these new things will behave.

For this reason, successful designers have learned to avoid the temptation to design everything in one big lump. Instead they may build one small part at a time, analyze the actual behavior, and then repeat the process for the next part. In this way, design becomes not only evolution-like, but actually evolutionary.

This process of incremental design takes place at two levels—in the mind of the designer and in reality. When it takes place in reality, it is *design as maintenance,* which is the principal mode of design today. The vast majority of design decisions actually put into effect today are created by maintenance programmers, not designers.

We do not mean to imply that this situation is a good thing—much of the "design" being done in maintenance can be equally well viewed as systematic deterioration. Think of a piece of farm equipment—a harvester for example. In the actual dust, rain, wind, and mud of a real farm, hundreds of tiny breakdowns occur in the life of such a machine. Each of them is repaired by the farmer with a piece of baling wire, or perhaps by a welder called in for a bigger break. Sometimes these repairs result in an improved machine—one that outperforms the "pure" machine as designed and delivered by the factory. In many cases the worker on the scene can spot in an instant a design flaw that the engineer, with years of technical training, could never have foreseen.

But at other times, the baling wire repair overlooks something the broader view of the original designer did not miss. Then, the succession of repairs eventually leads to the harvester becoming less effective. In the end, it has accumulated such a burden of

**103**

insults to its original design that it must be dragged off to die in the harvester's graveyard.

And so it is with computer systems. System designers would do well to spend a few weeks on a farm during the harvest. For the computer system can easily be visualized as a harvester of information. Great fields of data are fed into its jaws, the wheat of information is separated from the chaff of data, and the output is concentrated into a form one step closer to eventual human consumption.

In modern times, of course, farmers are more sophisticated, and farming operations are larger and more likely to support sophisticated repair operations. An elaborate network of repair services exists to support far more complicated farm equipment than was dreamed possible a generation ago. Americans seldom recognize the role of this network and knowledge. That's one reason Americans have been befuddled by their failure to export their agricultural technology. And, in a similar way, system designers fail to understand, or even to take into account, the maintenance environment in which their beautiful designs will have to survive.

The designer who does understand will not design to the fullest extent of her sophistication. She will resist the temptation to use her design as a vehicle for proving to the farmhands that her mind lives on a different—higher, to be sure—plane of existence. Instead, she will strive to keep the design transparent by basing it on familiar models; by using descriptive terms; by employing natural analogies; by acknowledging sources, rather than making new forms appear as if by magic; by knowing the level of the maintainers and users, and not going too far above it or below it; by developing and nurturing an *intuition* about the mind of the maintainer.

The typical agricultural engineer, even today—although trained to design agricultural implements in the university—was brought up on a farm. It is much more difficult for the city girl or boy to succeed in designing farm implements because they lack the intuition about the environment in which those implements will be used. Intuition is not magic. Intuition grows out of innumerable small and unconscious facts picked up from direct contact with reality. As agricultural engineering becomes more explicit, more sophisticated, and more scientific, it will also become further removed from the farm and the experiences of the farmer. In the process, engineers are likely to lose the very intuition that made their early designs such a success.

The same process, of course, permeates computing. Most system designers have been programmers. In their early design efforts, they succeed largely because of intuitive empathy with the programmer's plight. Then they start reading books and becoming more explicit, more sophisticated, and more scientific. Somewhere in the process they forget how comforting it is to a programmer to have a deck of cards, wrapped with a couple of thick rubber bands, snugly in hand. They forget how troublesome it is to search through a pile of listings to verify that the interface to a module is what it seems to be. They forget how tempting it is to patch in a change and leave the updating of the documentation until tomorrow, or another day.

As they advance professionally designers tend to forget their humble origins, the wellspring of their former success. Programmers who implement their designs cannot "think creatively on schedule" any more than the designer can. For successful design, there must be an *intuitive* link between the designer and the programmer, and between the designer and user.

You can go to college and learn a lot about design in the department of computer science, but the only department of intuition is in the school of hard knocks. If you're going to design a manure spreader, don't be afraid of starting with a little manure on your boots.

# The Goat and the Hippo: A Fable

*Early Thursday morning—or late Wednesday night, take your choice—*
*Las Vegas police got a call from a woman who said she had seen*
*a goat leading a hippopotamus down a highway. She was right. The*
*goat and the hippo had escaped from the Tulsa Spring Park Zoo and*
*were walking into the desert along the road. Officers enticed the*
*goat back to its pen. The hippo meekly followed.*

*International Herald Tribune*
Friday, July 27, 1973

Late one Wednesday night at the Tulsa Spring Park Zoo in Las Vegas, the goat noticed that the door to the main pen had been left unlatched. He thought of escaping, but realized that he wouldn't have much chance alone in the desert, so he waked his friend the hippopotamus. "Wake up, Harry. The gate is open and we can escape into the desert."

Harry blinked his six-inch eyelashes and tried to wake up enough to comprehend what George was saying. "Go into the desert? Whatever for?"

"To make our fortune, of course. We'll never amount to anything penned up in here."

"We'll amount to even less out there in the desert. A pile of bones, that's what we'll amount to. I prefer a little water, even if I have to be penned up to get it."

"But that's just it, Harry. Not very far from here, across a short stretch of desert, is Lake Mead. It's the lake behind Hoover Dam, and most of the time the lake isn't full, so there's miles and miles of glorious mud all around the edges."

"Really," said Harry, for the first time fully awake and paying attention. "But how could we make a living, if we were on our own?"

"That's simple," his goat friend continued. "Hoover Dam is one of the biggest tourist attractions around this part of the country. A lot bigger than this zoo, I'll tell you that. Now when the tourists get there, there's not really much for them to do—just look at the dam, and you can't do that for very long without being bored. Boredom is *the* fatal disease of tourism. So if we were there, we could put on a show for them and make a fortune."

"What kind of show? Neither of us has any talent, like my cousin Harriet, the Dancing Hippo."

"Talent? Who said anything about talent? Once you've got a captive audience of tourists crying out not to be bored, all you

106

have to have is something unusual—like the world's only upside-down tepee or the world's biggest collection of junkiest souvenirs . . ."

". . . or a goat and a hippo who are friends," finished Harry, who finally saw what George was driving at. "You've convinced me! Let's get going before the mud dries out!"

So off they went, the two friends, into the desert to start their new enterprise. But after they had walked about an hour, a lady who couldn't sleep saw them and called the police, who led them back to the zoo and locked the gate.

MORAL: *Not everybody shares your distinction between innovation and mischief.*

IN OTHER WORDS: People who understand your designs may not help you, but people who don't will inevitably get in your way.

# PART
# VI

# Tradeoffs

*The Water Moccasin and the Waterproof Moccasin:*
*A Fable (page 137)*

## Optimitis and the Tradeoff Concept

In the United States recently, much attention has been given to the prevention and cure of occupational diseases. We all know about coal miners and black lung disease, and some of us have heard that hatters in the nineteenth century were subject to mercury poisoning—whence the expression "mad as a hatter." But not many people know or care about the occupational diseases of computer people—except, perhaps, computer people themselves.

Many of these diseases have no known cure. The analyst will always have strained ear drums—from trying to understand muttering users. The programmer's eyesight will always be destroyed after a year or so—from reading too many manuals and dumps. Some diseases, like manager's ulcers—which are obtained from everything—can be reversed, but only by getting out of the business in time. But there is one DP occupational disease for which there is a cure, if only it were more widely recognized. The disease is endemic to designers, but known to all DP specialists. It is known as *optimitis.*

Optimitis is an inflammation of the optimization nerve, that part of the nervous system which responds to such requests as:

1. "We want this system to be produced at minimum cost."
2. "Give us the fastest algorithm."
3. "Get it done in the shortest possible time."

In a healthy individual, the optimization nerve receives such requests and sends an impulse to the mouth to respond, "What are you willing to sacrifice?" In the diseased individual, however, this neural pathway is interrupted, and the mouth utters some distorted phrases like "Yes, boss," or "Right away, boss."

The social cost of optimitis runs large. Anyone who has ever been stuck with a project conceived by a diseased designer will want to know about the cure—a kind of physical therapy using what I call "tradeoff charts."

Figure 5 is an example of a tradeoff chart that might be used to cure a designer whose problem is to "Design the world's fastest runner." The chart is a graph of speed versus distance for world records in running events. All tradeoff charts are graphs of this type—one performance measure versus another. What they show is how one performance measure has to be traded off, in the real world, against some other.

In this case speed has been traded against distance. Assuming that the world's record is about the best you can do at any given time, the curve of speed versus distance gives you something to shoot for in your design. It also gives you an idea of the nature of the relationship between these two measures of performance—a relationship that may well hold even for some newly designed runner.

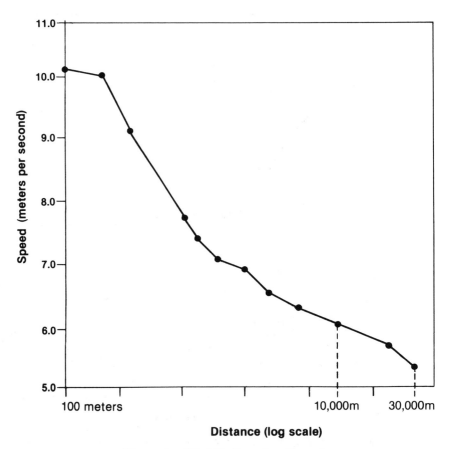

**Figure 5.** World's Running Records

What the tradeoff chart of Figure 5 says is that if you want to run faster, you'll have to restrict yourself to a shorter distance—assuming all other factors are held the same. Alternatively, it says that you can run farther if you're willing to go more slowly. But most important, it says that "You don't get nothin' for nothin'."

If someone asks you to run faster, you can offer to do so—if you don't have to keep it up for such a long distance. Or if a long-distance runner is needed, you may be able to get one—if you're willing to go more slowly. But you're unlikely to get a faster runner who can run farther as well, nor will you find a longer-distance runner who runs faster.

One of the reasons people are confused by optimitis is that they fail to recognize the limiting nature of the tradeoff chart. Figure 6 shows a plot of speed versus distance for a particular

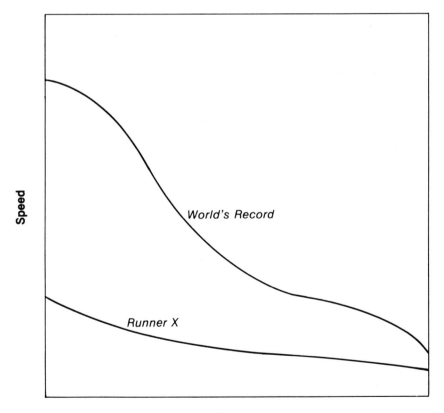

**Figure 6.** Performance of a Long Distance Runner Compared with World Record Performances

runner—not a world record holder at any distance. Because the tradeoff chart is composed of world records, we know that runner X's plot will *never* surpass it. Runner X's curve represents a particular "design" relative to the "best possible" design on these two dimensions. Reading the curve, we can characterize what sort of runner X is—a slow starter and not much of a sprinter, but with good endurance at the long distances.

In Figure 7 we see another curve for runner Y, who might be characterized in words as "a sprinter who cannot go the distance." In Figure 8 we see my own curve, which in all modesty I might describe as "a lousy runner at all distances."

Seeing all these figures, we begin to understand how the tradeoff chart cures optimitis. When someone asks the diseased

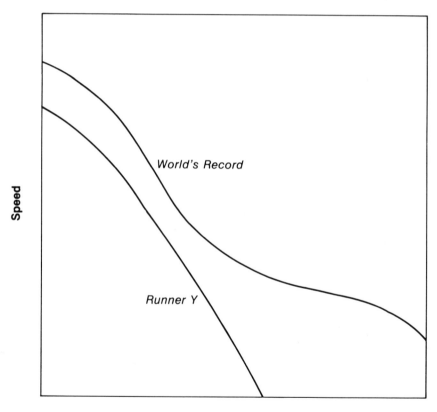

**Distance**

**Figure 7.** Performance of a Sprinter Compared with World Record Performances

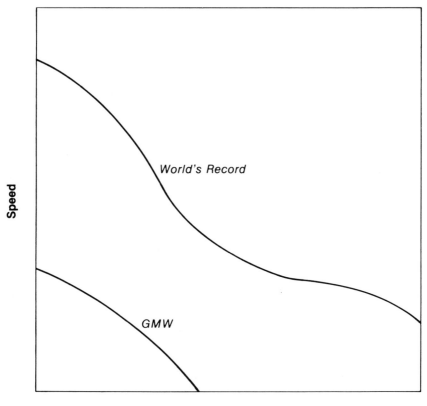

**Figure 8.** Comparison of Gerald M. Weinberg's Performance with World Record Performances

designer to "design the fastest runner," conditioning through long therapy must trigger the reaction: "Let me check my tradeoff chart." After creating the chart, the designer will naturally think to ask such questions or make such statements as:

1. At what distance do you wish to work?
2. It will be relatively easy to get a system that runs fast at 100 meters, if it doesn't ever have to run 30,000 meters very well.
3. We can probably find an off-the-shelf system, like GMW, a lot easier, which will lower the cost, but if you want world-record performance, there aren't going to be many candidates.
4. If you need long distances, your 100 meter times may not be so good.

Once the questions start, there's a chance that the design effort will be on a reasonable path, acknowledging that you don't get anything without paying for it.

The converse, of course, isn't true at all. You may very well pay for something and not get it. For example, you may hire GMW at professional athlete rates, but even a million dollars won't get him running 100 meters in under fifteen seconds. The healthy designer spends time trying to get a healthy curve, not an impossible one.

There are many tradeoff charts in the DP business. Programmers trade off space and time, access time and capacity, coding effort and testing effort. Designers trade implementation cost and operation cost, flexibility of use and ease of maintenance, functionality and implementation time. Analysts trade one user's satisfaction for another's, conformity to standards for conformity to present convenience, precision of specification for time to implement. Managers trade experience for cost of personnel, hardware costs for software, budget for morale.

Despite the diversity of tradeoffs, all can be reduced, to a first approximation, to a tradeoff chart that's shaped like Figure 5. The labels are different. The scales are different. But the thinking is the same: *Moving in one direction incurs a cost in the other.* Until you master the art of thinking in tradeoff terms (and then learn to juggle simultaneous tradeoffs), you'll remain a technician, rather than a professional.

## Tradeoffs—Quality versus Cost

One tradeoff we find in almost any system design or implementation is that of *quality* versus *cost*. The generalized tradeoff curve of Figure 9 could apply to any of these situations. For purposes of discussion, let's first assume that:

1. Quality here means absence of bugs in a delivered system.
2. Cost here means the development cost to deliver that system.

Just what does the tradeoff curve represent in this instance?

As usual, the tradeoff curve represents the best we can do under the present circumstances, all other factors being held equal.

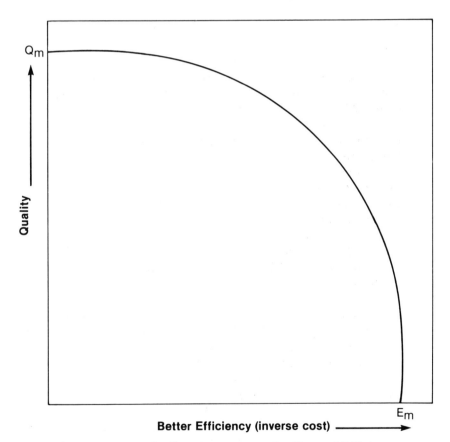

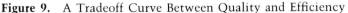

**Figure 9.** A Tradeoff Curve Between Quality and Efficiency

117

We can't get better quality than $Q_m$, no matter how much we're willing to spend. Nor can we increase our efficiency (decrease our cost) above $E_m$, no matter how poor a system we're willing to accept. In this case the limiting cost must be zero, if we're willing to accept a system that does nothing right. If there's an existing system that is satisfactory, zero cost could represent better than zero quality—just keep what we have. That's the strategy so many installations have adopted in the face of the uncertainties of development.

How do these uncertainties relate to the tradeoff curve? The tradeoff curve only represents the limits on what we can do. We can do worse—anywhere in fact, within the closed portion of the curve. The curve represents what we are capable of doing. Only events will say what we will actually do.

In a well-managed organization, with a little bit of luck, we can choose any point along the tradeoff curve as our target for a new system development. If we want to allow for bad luck or poor management, we can even choose a point *within* the curve. What we cannot do—without changing our techniques of development—is choose a point *outside* the curve.

Let's see what this means. We've been producing systems that slip into production with about one error per twenty lines of code—let's say that is our measure of quality. These systems cost us about $50 per line of code, or twenty lines for $1,000—our measure of efficiency of production. This point is labeled P on Figure 10.

Our users begin to complain about the junk we're delivering, arguing that the errors are costing them enormous sums to rectify in production. They ask that the number of errors be cut in half—to one in forty lines of code. If we know our quality/efficiency tradeoff curve with any precision, we may be able to tell them just where this new point, R, lies, and what it will cost them, $E_R$, X, to attain it. We will also be able to forestall any foolish requests such as:

1. "Since you can give us twice the quality for thirty percent increase in cost (R at $E_R$), why don't you just give us four times the quality (S) at sixty percent increase in cost?" Reply: Because of the shape of the curve, near point P, additional quality is going to have a disproportionate cost, such as $E_S$.
2. "Why don't you just reduce the errors to zero?" Reply: Because the curve is competely flat as we get close to zero errors, meaning that no amount of money is going to guarantee getting the last error out before delivery.
3. "Well, then, just lower the quality a bit and cut our price in

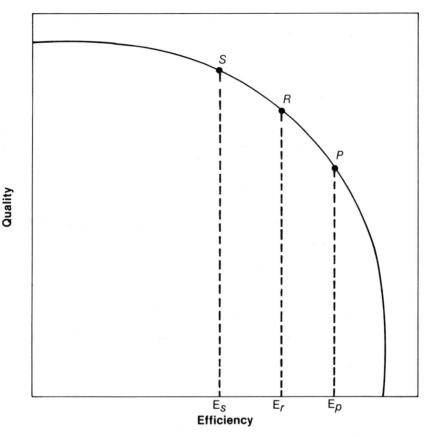

**Figure 10.** Three Different Operating Points with Different Choices Between Quality and Efficiency

half!" Reply: There would be very little savings from dropping the checks in our development process, because of the shape of the curve to the right of $E_p$. We could probably cut ten percent off our price if you're willing to take ten times the number of errors.

In other words, we can turn generalized grumbling into specific alternatives for action. Of course, the actual tradeoff curve for our installation won't be known with perfect precision, for a variety of reasons:

1. The curve requires "other things being equal," and other things are never equal—people differ, problems differ, and there is such a thing as bad luck and good luck.

2. We won't have measured things that well in the past, if we're a typical installation.
3. Our abilities change with time, so the curve can change.

But the curve gives us a starting point, even to discuss the effects of change. Indeed, a particular curve can be considered to represent the entire "way of doing things" that our installation calls system development, including experience, training, tools, management, procedures, and even expectations. If we change our "way of doing things"—our technology—we will pass to a new curve.

For instance, suppose we introduce some new combination of programming practices, such as formal reviews and a development library. If these are "good" practices, they may move us to an entirely new curve as shown in Figure 11. This curve lies above the previous technology curve in some places, but below it in others. Why? Because the new practices represent added costs directed toward improving quality, they may prove counterproductive if we are not interested in high levels of quality—the part of the curve labeled X. Conversely, these particular practices may allow us to achieve levels of quality we could not have attained with the old technology no matter how much we were willing to spend—the part of the curve marked Y.

More generally, however, as the curves suggest, a new technology represents *choices* of how much quality we'll trade for how much cost. Because of the shift to the new practices, these choices are not always explicitly or wisely made. They are not an automatic consequence of the new practices, so we may be disappointed if we just let nature take its course, rather than seize control of our own destiny through the period of change.

For instance, in moving to the new technology from point P on the old technology, we could choose to:

1. Keep costs constant, while improving quality (Q)
2. Keep quality constant, while reducing costs (E)
3. Improve the quality somewhat, but also decrease costs somewhat (C)
4. Improve quality at the same time we're raising costs (QQ)
5. Cut costs, now that we can do it without slipping quality too much (EE)
6. Let everything go to pieces (X).

If we fail to set explicit quality and cost goals before embarking on a new technology program, the direction we move from

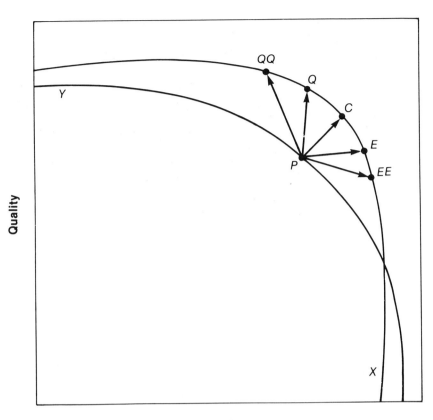

**Figure 11.** Five Different Choices of Moving from the Tradeoff Curve of One Technology to the Tradeoff Curve of Another Technology

our present position, P, is likely to be random. Thus, even though we should be on a "better" curve, we may find ourselves less satisfied with the result. This effect doesn't explain all disenchantment with new technologies, but it does explain many of them, such as:

1. Adopting a new programming language and finding that "costs have increased" (because the programmers are now setting higher standards of quality)
2. Adopting a new programming language and finding that "quality has dropped" (because the programmers are using the new language's facilities for quick and dirty solutions)
3. Adopting a new programming language and finding that "we haven't saved as much as we thought we would," and "the

new programs aren't that much more reliable" (because in the push and shove we've wound up at point C, which happens to satisfy nobody's preconceptions of where we were going).

Remember, we've just been discussing quality versus cost in program development. The same type of curve can be drawn for many other tradeoff situations, including many other quality/cost tradeoffs. But for now, try to sketch the curve for your installation's present technology. Then sketch the technology you had a year ago, and the one you hope to have next year. Finally, draw the point you're now operating at (P), and the points you came from and are (hopefully) heading toward. I think you'll find it a quick and instructive exercise, especially if you involve a few friends.

# Trading Analysis for Design

*All government, indeed, every human benefit and enjoyment, every virtue, and every prudent act, is founded on compromise and barter. We balance inconvenience; we give and take—we remit some rights that we may enjoy others . . . Man acts from motive relative to his interests; and not on metaphysical speculations.*

Edmund Burke

Because tradeoffs are universal, we can often use the tradeoff concept to reason backwards from effects to causes. The reasoning process goes like this:

1. We are doing X, instead of Y.
2. We must be getting a benefit from X—what is it?
3. We must be losing something from not doing Y—what is it?

Here's a concrete example. Some organizations don't separate the work of systems analyst and systems designer, while others do. What benefit is gained from making the separation? What benefit might be lost? By examining the evolution of systems analysis, we notice the tendency to separate analysis and design as systems grow larger and more complex. This correlation suggests that the separation is another example of "divide and conquer" to beat the Square Law of Computation.

But what was the compromise? What did we trade for this simplification of an overly complex system development process? Obviously we pay some extra "communication" costs, and may incur problems if communication slips somewhere between user and designer. But by having an analyst who is a specialist in communication, we may actually do a better communication job because of the separation.

What else could it be? Let's reduce the problem to a down-to-earth example—the design of a house. If you were asked to specify your ideal house, you might start by saying "I'd like forty rooms." Now that's a pretty clear specification, and any architect can work that desire into a design. The problem is that this house is going to cost you a lot of money—to build and to operate. Even though the architect may meet your specification exactly, you may be very unhappy with the design.

Real-life architects seldom operate this way. As you specify what you want, they are constantly sketching rough designs in their heads, evaluating the costs and feeding back rough estimates. As you hear these estimates, you make compromises with your initial desires, so that the final specification reflects both a specification process and a design process.

Some people approach the architect differently. They would really like a ten-room house, but for some reason think they can't afford one. When the architect asks them how much space they want, they compromise in their heads and come out with "How about six rooms?" Architects aren't mind readers, so they proceed to design a house with six rooms. And even though they can't read minds, they seem to be able to read purses, so the cost of the six-room design just matches your six-room figure bank balance. You might have got a ten-room house of less lavish construction, but you never asked for it. You'll never know what might have been.

Few people can specify exactly what they want in a house, if a cost factor has to be included. Because they aren't architects, they can't know what tradeoffs are realistic, and even an architect can't know without some extensive design work. In architecture, specification and design go hand in hand, making the architect's life more complex, and making the actual process more costly.

Different types of architectural work traditionally operate with different degrees of separation between specification and design. We might plot the architect's fees as a function of this separation, as shown in Figure 12. On the right (separation = 1), we have mobile homes in which the only choice the buyer gets is color. Moving left we might have tract houses, then custom houses, then all sorts of special buildings like offices and churches. At the far left we might have cathedrals, where the job is being respecified for 300 years while the actual construction is going on. The other side of this curve might look like Figure 13. As the separation grows, there is less and less chance for the customer to compromise one thing that's desired for another that's desired a bit less. The designer reads the written specification and tries to guess how much floor space the customer would be willing to trade for a stone fireplace, or how much processing cost for one second faster response time.

To some extent, the problem can be alleviated by good communication between the designer and the customer. But with a separate analyst between the two, this communication can be difficult to accomplish without offending the analyst. If you keep the communication formal, you engage in an endless series of user-to-analyst-to-designer-to-analyst-to-user cycles.

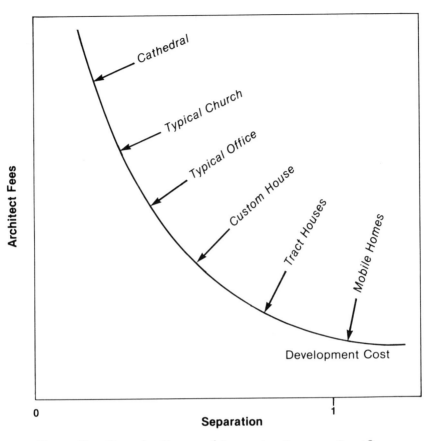

**Figure 12.** How the Degree of Separation Between Specification and Design Affects Architect Fees

To avoid these problems, whether or not there is a separate analyst, you must relax the formality of the entire specification process. If a user has been forced to be extremely formal and detailed in specifying what is wanted, then that user is not predisposed to compromise later when design considerations dictate compromise. It's only human to resist compromising our interests, so if we are to foster successful compromise we must make it as easy as possible for people. The more committed we become to what we want, the less likely we are to compromise. Although we need "tight" specifications for successful system development, users rightly expect that tightness should apply both ways. If they have to accept what

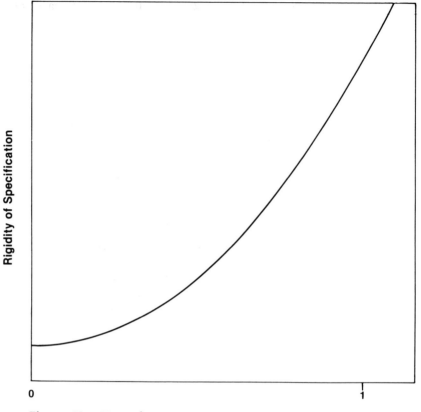

Rigidity of Specification

0                                                        1

**Figure 13.** How the Separation Between Specification and Design Affects the Rigidity of the Specifications

they originally specified, why should they accept compromises suggested by the designer?

In *The Golden Notebook*, Doris Lessing observed:

> It seems to me like this. It's not a terrible thing—I mean it may be terrible, but it's not damaging, it's not poisoning to do without something one really wants ... What's terrible is to pretend that the second-rate is first-rate. To pretend that you don't need love when you do; or you like your work when you know quite well you're capable of better.

We don't have to get everything we want in our systems to be happy with them, but we do need the chance to make our own choices about what's important to us and what's not. When specification and design get too rigidly separated, users pass control of their compromises to the analyst and designer, where they don't really belong. Sometimes users complain bitterly about the compromises that have been made "on their behalf." That's bad enough, but even worse are those users who simply accept such loss of control as an inevitable part of "the way systems have to be developed." They say nothing about their dissatisfaction, or blame themselves for "not knowing what I wanted."

But worst of all, in my opinion, is when they "pretend that the second-rate is first-rate"—that what they got is what they wanted all along, even if they didn't realize it. For without some candid feedback from users, analysts and designers are never going to find the right tradeoff between analysis and design.

# A Tradeoff View of Error Correction

I've written a lot about error correction in general and spelling correction in particular. (See, for instance, *Humanized Input*, which I wrote with Tom Gilb.) Now that spelling correction hardware and software are appearing on the market, it's time to make design decisions on the subject. The first of those decisions has to be this: Should we attempt to correct spelling?

This is the first question because hardly anybody thinks of it as a question at all. Naturally we want to correct spelling—don't we? Spelling errors are bad—aren't they? Correct spelling is good—isn't it? How can anyone who went through umpteen years of school have missed the message?

Because everyone blindly accepts spelling correction as "good," it's an excellent example to use in illustrating the fight against optimitis. Let's have a look at the entire subject from the point of view of tradeoffs. Surely we can spend a few minutes thinking before rushing off to spend our money or write programs. Isn't that extra hesitation the essence of design? With a need like spelling correction, which "everybody understands," you have to be particularly careful. Spend a lot of time brainstorming, alone and with others, about just what you expect spelling correction to do and not to do for you. Here's an example from yesterday's editorial work:

> "Their enthusiasm was transformed from high to low by chance of events".

When I read that, I knew it didn't quite make sense, but what was the spelling error? Sally thought that the word "chance" should have been "change." I thought that the word "of" should have been "or." What would the spelling corrector have done?

The answer, in this case, is probably "nothing." Both "chance" and "of" are legitimate words. Thereby hangs the first and foremost limitation of any spelling correction system, whether it be people or machines: *If the mistake makes sense, it probably won't be recognized as a mistake.*

I've kept records of proofreading on many of my books and inevitably found that the errors that persist longest take the form of words that are transformed into other words. As Oscar Wilde once remarked, "A poet can survive anything but a misprint." In poetry the sense is often difficult to determine without careful reading, and the transformation from "chance" to "change" could easily produce a perfectly acceptable—but different—poem.

In business documents, such as contracts, an error of this type can be similarly fatal. One client told me of an overseas telegram that answered the question: "Should we purchase additional storage devices or relocate some devices from other installations?" The reply was something like "More storage devices," but it should have been "Move storage devices." Nobody suspected that there had been an error, so there was no possibility of correction.

Clearly, no spelling correction system will be able to detect, let alone correct, every typographical or spelling error. Nevertheless, the very term, *spelling correction,* may lull people into overconfidence. This, then, is the second limitation: *Always examine the possibility that improved spelling may lead to excessive confidence in the text.*

Certainly we computer people are familiar with the problem of overconfidence in computer output. I've written before about the "Titanic Effect," where you don't supply sufficient lifeboats because the ship is "unsinkable." Be sure—be doubly sure—that you're not creating a Titanic by putting spelling correction into some system. Ask and ask again. "Are we dropping other safeguards against the consequences of error to help pay for the spelling correction? Will less frequent errors slipping through the system actually increase the cost of error so much that the total cost of errors will actually increase?"

Costs of undetected errors can increase, for example, when people become unfamiliar with procedures for handling such errors when they finally are detected (Figure 14). This was a common experience when hardware became more reliable. Field engineers had fewer chances to learn how to handle errors, for instance, when vacuum tubes were replaced by transistors.

Another reason the cost of errors might increase is not because of problems fixing them, but because of the trouble they cause when they do slip through. We may, in fact, bet more on the system producing "correct" output. Don't ask merely, "Can our system *stand* an undetected error?" It had damn well better be able to stand it, for there are sure to be some. Ask instead, "How does the cost picture change, under spelling correction, for undetected errors?" You'll probably find a curve like that in Figure 15, with a central region in which the net cost of errors is smaller than it would be with more errors or fewer errors.

Another important factor to consider in drawing these trade-off curves is the possibility of the spelling corrector actually making matters worse. Whether it's a person or a machine, there will inevitably come a time when the "correction" is worse than the original.

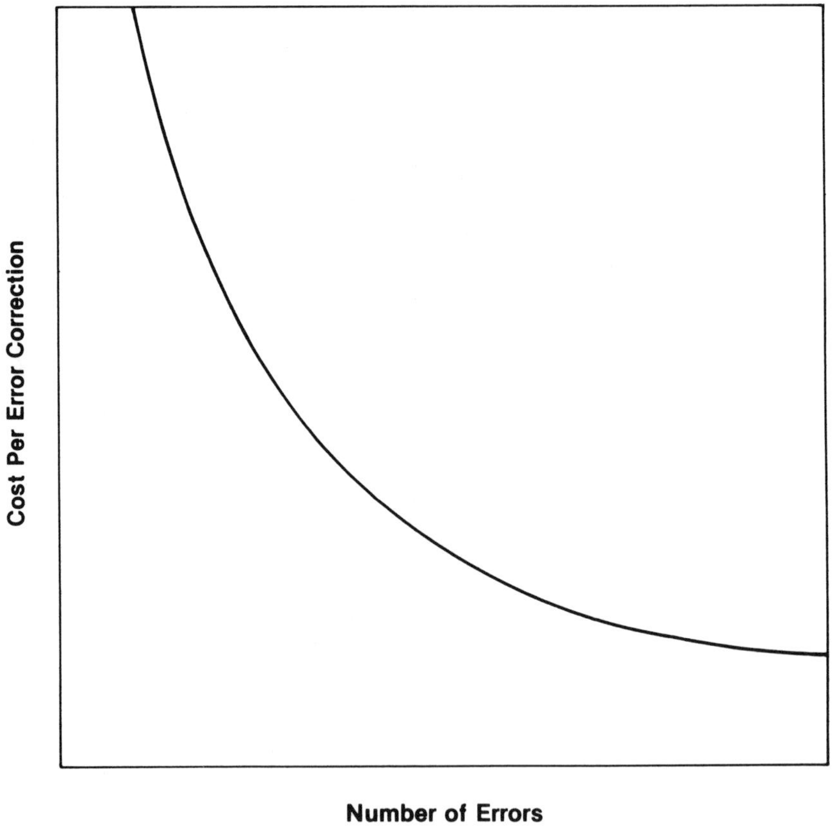

**Figure 14.** The Cost of Correcting Each Error May Rise as
there are Fewer Errors to Correct

As an author writing about errors, I face this problem all the time.
Once, when writing about a spelling correction algorithm, I had an
editor who corrected the spelling in each input example, causing
repeated nonsense such as *significant* being corrected by the program to *significant*. Readers would go over and over such a sentence
trying to figure out what it was talking about. (If the above "corrected correction" slips through this editor, you might understand
what I'm saying, but there's no way to guarantee that they'll leave
it alone.)

The best technique I've been able to come up with is to put
all "intentional" errors into figures which I supply separate from
the text. But even there, the helpful correctors often do me in. In
*Structured Programming in PL/C*, we were trying to show a bug

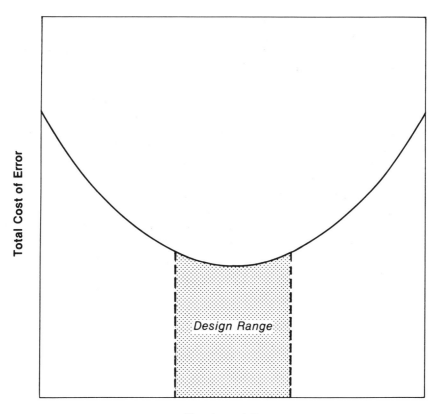

**Figure 15.** Total Cost of Error Handling may be Smallest at Some Finite Rate of Errors

in a program that printed a calendar. The bug caused a page skip in the middle of the calendar, as the figure clearly showed. But just before sending it to the printer, the helpful production people—using razor blades, glue, and infinite care—cut out the gap and pieced the calendar together "correctly." Alas, tens of thousands of readers were utterly baffled by the entire discussion of this bug that never appeared in the book.

Intentional errors are not merely problems for authors. Ask yourself, "Will we be able to slip unusual cases through the watchful eye of the spelling corrector?" For instance, could it handle correspondence with our clients in the colonies who prefer to read 'color' and 'honor' instead of 'colour' and 'honour'? (Will that sentence make it intact through the editorial process?) As I was writing this, Crystal came in with a question about the mailing list she was creating: "Should this person, 'Jin Shyr,' really be 'Jim Shyr'?" Fortunately she had the sense to ask, but how would an automatic system have handled it? (The answer, incidentally, was no. "Jin" was correct.)

What all this amounts to, really, is a plea to all my colleagues out there to try to do better this time than we did the last. Spelling correction has great promise, but let's not promise too much, or too soon, as we usually do in the computer business. Don't go into spelling correction with a big advertising splash among your corporate executives. Try an incremental approach for a change. Get a software package, by all means, if the price is low enough to allow you to fail without embarrassment. Or program some simple correction scheme and observe it in action until you learn more about the tradeoffs of correction in your own environment. Don't stand in the way of progress, but be a bit skeptical. And humble. (Or was that "bumble"?)

# A Cribbage Lesson

Once upon a time, I subscribed to a magazine called *Organic Gardening*. For the first year, each issue was a revelation, but when the second year rolled around, I began to read the same articles, gardening tips, and even advertisements for the second time. As it turns out, there are just so many ways to grow a potato—even organically. So I just saved the first year's issues and cancelled my subscription.

I'm beginning to feel the same way about certain "personal computing" magazines. Not wanting to be old-fashioned, I subscribed to everything available on small computers, even though many of the publications were destined to fold before I got my subscribed number of copies. Now I'm down to a small number of survivors, but I'm still getting a strong sense of déjà vu.

Actually, I'm getting déjà vu on two levels. First, each issue seems pretty much like the previous issue, with "how to implement strings in XBASIC," "How I balance my personal checkbook with my PET," "Math puzzles that will amaze you," "All the problems I had with my off-brand hardware until I saw the light," and "Krazy Komputer Kartoons" that present the computer as some sort of humanoid.

But among all this froth, you find another level of repetition—not from last month or last year, but from twenty-five years ago. And that's worth the price of a subscription—to feel young again. Issue after issue, they're making the same mistakes I made when I was their age. The only differences I can see are these:

1. Their mistakes don't cost so much.
2. Their mistakes are published.

The precipitating cause of this column was the following letter to a personal computing magazine from a contributor who had published a CRIBBAGE program the previous month. I had glanced at the program wistfully because over twenty years ago I had implemented a cribbage playing program on an IBM 650. At that time I thought nobody would be interested, so I kept it to myself. Besides, I think my employer was happier that way.

But nowadays, when you write a cribbage playing program, it's a world-shaking event, and must be published for all to see. So here's the letter:

> Several readers have contacted me to report that my CRIBBAGE program will actually stoop so low as to cheat during the play of the

hand. The computer will on occasion, after a "GO" situation, replay a card it has already played . . .

I will also mention that there will be a problem in the unlikely event the computer has four 5's and has to play first; it has a rule to never play a 5 as the first card.

My apologies to those who might have spent frustrating hours searching for errors in the input of the program to their computer systems or trying to understand the logic of the program.

First off, the letter made me feel very smart, because I had glanced at the program and decided that if for some reason I wanted to play cribbage with my computer, I'd find it easier to program it myself than to copy CRIBBAGE. On the other hand, I lost the opportunity to catch my machine cheating—something I've always wanted to do.

Actually, I have a theory, tested by much empirical evidence, that every computer game published in these magazines has at least one error and one failure in the case of an "unlikely event" that's bound to happen. Indeed, my theory goes further. If the games didn't have these errors, they wouldn't be nearly as much fun. After all, it's easy enough to find a human being for a cribbage partner, and often more rewarding. But no matter how tempted you might be, you generally aren't allowed to debug another person.

And there, I think, lies the essential difference between the "professional" and the "hobbyist." Although they both use computers and write programs, one wants to get a program that does the job and the other is hoping that the program will fail in some challenging way. (By this definition, of course many highly paid programmers will have to be classified as hobbyists.)

The difference can be seen clearly if we rewrite the letter as if it came from a programmer in a software firm. Receiving the letter are 10,000 companies that were using an accounts receivable package:

Several clients have contacted me to report that my RECEIVABLES program will actually stoop so low as to cheat during customer billing. The computer will on occasion, after issuing a bill, rebill an item that has already been billed and paid for . . .

I will also mention that there will be a problem in the unlikely event that some bill ever has four charges for the same type of merchandise or service; it has a rule never to issue a bill with four duplicate line items.

My apologies to those who might have spent frustrating hours searching for errors in the input of the program to their computer systems or trying to understand the logic of the program.

One can anticipate approximately 9,437 telegrams reading:

Apologies be damned! We're suing you for lost customers and lost billings to the amount of $47,980.

That's the difference between the professional and the hobbyist—the lawsuits, or even worse, hanging over your head with every dropped period or misplaced branch.

It's easy to dismiss the hobbyists as frivolous and trivial, because there are essentially no consequences whatsoever to hobbyist programming mistakes. And because there are no consequences, there is damnably little incentive to produce good clean code. But there's an important insight here which would-be professionals might miss in their haste to scorn the hobbyist.

We know that without pressure, performance can be quite minimal, but psychologists also tell us that too much pressure also has a deleterious effect on performance. The classical pressure/performance curve is another tradeoff, something like that of Figure 16. Performance crashes mightily when the pressure gets to be too much for a soul or organization to bear. We sneer at the hobbyist for living on the left-hand slope of this curve. But doesn't it worry you just a bit that all this talk of lawsuits and worse may have pushed many of our brothers and sisters over the top and down the right-hand slope?

All these years we've seen generations of programmers and machines come and go, with the same mistakes and silly games being played on and on by each new generation. But something *is* changing all this while. Some systems are getting bigger, more complex, and much more critical if they fail. In other words, our chances of success are smaller, and the consequences of failure are greater.

As I travel about and talk to our students, I'm hearing an increasing number of stories about what's happening to people who were pushed over the line. I think the problem has become serious enough for all of us to start thinking of solutions. Interestingly enough, the most common tactic among individual programmers under pressure seems to be to get a personal computer, take it home, and program it to play games. What better safety valve for the pressure of having to write programs that play the game according to the rules?

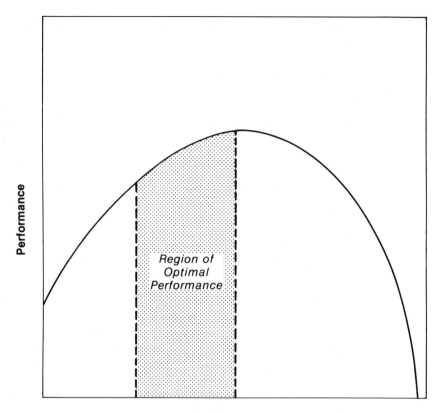

**Figure 16.** How Pressure and Human Performance are Traded

# The Water Moccasin and the Waterproof Moccasin: A Fable

While running through the swamp one day, a little boy lost one of his new waterproof moccasins. Though he searched and searched, he could not find it, so he went home for dinner limping and hopping on one foot. The moccasin, meanwhile, had floated some distance away, until it got caught among some branches. While it rested there, thinking of what to do next, a long snake came swimming up to it.

"You're new around here, aren't you?" asked the snake with an eye to getting acquainted. To tell the truth, the snake did not have too many friends in the swamp, and he liked to be the first to greet new arrivals, before the others started telling bad gossip about him. He was actually a very friendly snake, and the gossip was cruel and unjustified, but to most of the inhabitants of the swamp, a snake was a snake and that was all there was to it.

"Well, yes," replied the moccasin. "In fact, I just arrived this afternoon." He was a little shy about speaking to strangers, having passed all of his short life in the company of his twin, the right-hand moccasin.

"Then let me welcome you to our swamp. I'd shake your hand," offered the snake, "but as you can see, I don't have any hands, as I am a moccasin."

"Hey, really!" exclaimed the shoe. "But I'm a moccasin, too!"

"That's funny, you don't *look* like a moccasin."

"I was just going to say the same thing about you. Are you sure you're a moccasin?"

"Of course I'm sure. I may not be very smart, but I've been a moccasin—a water moccasin—all my life, that much I know."

"Oh, well," said the shoe. "That explains it. You're a water moccasin, but I'm a water*proof* moccasin. Guaranteed never to leak on the rainiest days, and lifetime plastigranite sole to boot."

"How interesting. Well, there aren't many moccasins of any kind around here anymore—and as you know, some creatures are a bit prejudiced against us. So we moccasins have to stick together. Why don't you swim over with me to my house for a bite to eat? Are you hungry?"

"I don't think so," said the waterproof moccasin, mainly because he didn't know what "hungry" was. "But I would like to come to your house—only I can't."

"Why not?" The water moccasin had been refused before, but never by a fellow moccasin.

The shoe sensed his suspicious tone and hastened to reassure his new—his *only*—friend from the swamp. "Not because I don't want to. But I'm stuck here in these branches and can't get out."

"Don't be silly," laughed the snake. "Just wriggle out and come along with me."

"What's 'wriggle'?"

"It's like this," the snake answered, giving a short demonstration that created a series of ripples in the water. The ripples made the shoe bob up and down, which the snake took as his efforts to wriggle, but he didn't get loose. "That's the way," he encouraged, "but you'll need more action, more flexibility, if you want to get loose."

"But I can't," said the waterproof moccasin, who was beginning to cry at the prospect of spending the whole dark night alone in the swamp before someone came back to find him. "I can't."

"Well, I never! What kind of a moccasin are you if you can't wriggle any more than that?"

"I guess we waterproof moccasins weren't made to wriggle the way you water moccasins can. I think it's this lifetime sole of mine. My sides are flexible enough, but it's so stiff it won't let me wriggle."

"Then what good is it?"

"I don't know, I guess. Until today, I used to think it was good to have a lifetime sole. Now I'm not sure."

"Well, *I'm* sure," answered the water moccasin as he began to swim away. "What good is anything that lasts a lifetime, but binds you so tight you can't even have fun. I'll see you around."

And so the waterproof moccasin remained tangled in the branches until one day it rained and he filled with water and sank to the bottom of the swamp, never to be seen again.

MORAL: *Never trade your immortal soul for a lifetime sole.*

IN OTHER WORDS: If you try to trade off flexibility for long life, you may wind up losing the long life, because no "lifetime" design can anticipate everything.

# PART
# VII

# The Designer's Mind

*The Confusion Technique: "Once you understand the essential function of the drive-in, you graduate to the question of selecting movies." (page 153)*

# Design as a Human Activity

*Man is but a reed, the most feeble thing in nature; but he is a thinking reed.*

Blaise Pascal

Natural selection provides one model of design, but there are many others as well. One of these is so simple and helpful that it's hard to believe so many writers on the subject have failed to mention it. Whatever else design may be, *design is something done by human beings.* By and large, design is a mental process of human beings—a kind of thinking. Nevertheless, we dare not forget the physical processes that can help or hinder the mental ones of such feeble reeds as people. Design, like the Zen assembly of a motorcycle, begins with peace of mind, but peace of mind can never be achieved by ignoring the very real demands and weaknesses of the body. There is no sense setting down design principles that real flesh-and-blood people cannot possibly act upon. Principles of design have to be robust, like a good motorcycle, in that they work even when the road has a few potholes, or the rider isn't Evel Knievel. And like motorcycles, or any other human device, design principles will sometimes fail even on the best road with the most experienced rider. If the success of design were not problematical, anyone could design, and designers would be impoverished.

To emphasize design as a human activity is to reveal some of the limitations inherent in design. Where human capacities are limited, so must design be limited—or we risk failure. We must not design systems that require people to do what people cannot do—that much is obvious. But do we obey this principle in the design process itself?

The designer is human, and the human mind is limited in its ability to predict the behavior of an unbuilt system. To overcome this limitation, we often aid the mind with analytical tools, such as diagrams and equations. But more often we use the opposite approach, often unconsciously. Instead of increasing the power of our mind, we decrease the complexity of the problem. Instead of

**141**

considering all potential variations of a design, we omit certain cases because we cannot analyze them with our limited minds and tools.

A striking and somewhat curious example of such simplification is found in systems large enough for statistical analysis. If we try to design an on-line personal computing service for three or a dozen users, we have great difficulty predicting performance characteristics such as response time. But if we design such a system for three hundred or a thousand users, we can effectively model the user behavior by a statistical distribution function. Then we have all the apparatus of statistics to help us predict, often with uncanny accuracy, how the unbuilt system will behave—statistically.

Part of this simplification of large number systems comes from a relaxed requirement that may pass unnoticed. When we are using a large system, we expect to be served more or less statistically. On the other hand, when we know there are only two other users, we expect to be treated with some sort of personalized service. Imagine dining in a restaurant where the lone waiter was spending all his time serving the other two tables. We may assume that he is playing favorites among the customers, and become agitated, but perhaps he is merely incompetent. The system designer, of course, must be enough of a student of human behavior to anticipate such psychological effects, and to account for them in the design.

The more obvious simplification of large number systems comes, paradoxically, from their very *randomness*. We take advantage of the assumed *independence* of the users' behaviors in creating our analytical models. At times, designers forget this independence assumption, or overlook a possible dependency. Before World War II, telephone systems in the United States were designed to rely on the fact that people used their phones more or less at random. Although there were many subscribers, there would be no reason for a large number of them to pick up the phones simultaneously. This simplification allowed the system to be designed with much less equipment in the central office.

Old telephone engineers tell how the entire system crashed the day Franklin Delano Roosevelt died in 1945. Many people in the country heard the news on the radio and went straight to the phone to tell somebody the news. At the time of designing the original telephone approach, there had been no significant radio broadcasting and the engineers failed to imagine an event that could destroy their simplifying assumption of independence in phone behavior.

By being more aware of their own behavior as designers, the telephone engineers might have avoided this problem, which could have had frightening results. (There was a war on!) The designers believed that independence was necessary to simplify the *system*, but it was necessary to simplify *their thought processes*. Had they been aware of the source of the problem, they might have realized that they could not possibly assume independence all the time. What if there had been a supernova, visible to all in bright daylight? What about a major earthquake—certainly no time for the phone system to collapse?

The self-conscious designer must anticipate such dependencies. Where dependencies exist, the designer can sometimes treat them as special cases. In the telephone network as redesigned, a sudden dramatic overload will send the system into a planned graceful degradation, rather than a collapse. Under ordinary circumstances, no subscriber will ever notice the difference between the two designs, but the difference exists.

A radically different approach is to have the system itself make the users notice. By doing so, the designer lets the system create the independence it needs to meet its performance goals. The telephone company charges variable rates depending on the time of day. In an on-line computer system, we can use the system itself to issue constant reminders to ensure that users know when they are burning up very expensive time.

Most designers would consider the scheme of variable charges to be a device for allocating costs more fairly among the users—those who use the more expensive facilities are to be charged accordingly, and facilities used in the state of maximum load are by definition expensive. Some few designers have a broadened perspective, and see the variable charges as a device for *shaping the behavior of the users* to suit the limitations of the system. But only a very few designers comprehend this scheme in its fullest sense—a device for *simplifying the mental task of the designer*!

# Design—The Reality and the Romance

> *Women are always blamed because they are fickle, but I sympathize*
> *with them. Some call it a vice, but to me it seems a vital necessity.*
> *If a lover is disillusioned, he should not blame her, but his own*
> *ignorance, since whether they are young, old, pretty, or ugly, they*
> *all do it!*
>
> Don Alphonso, in Mozart's Cosi Fan Tutte

Ferrando and Guglielmo, Don Alphonso's two young friends, have just lost a wager on the fidelity of their lady friends. Although he requires an entire opera to teach them this lesson, Don Alphonso's hundred sequins were never in serious danger. Young and romantic, the naive soldiers were unable to distinguish the behavior of flesh-and-blood women from some peculiarly idealized romantic notions of fidelity. Ferrando and Guglielmo are crushed by the heavy hand of reality, but Don Alphonso consoles them with some excellent advice:

> Take them as they are. Nature can't be expected to manufacture two unusual ladies specially to match your pretty faces.

Watching the opera, I fantasized Ferrando and Guglielmo in the role of young systems designers, enamored of the latest fashion in purified, idealized design. I rather identified with Don Alphonso, the dirty old man, heavy with experience, yet unable to convince his young friends without a demonstration. I decided to write a short lecture on my philosophy of design.

Like Don Alphonso, I believe in starting with an attempt to understand and accept the state of the world today—*the way things are*, rather than the way I would like them to be. Not that I try to know exactly how things are—that's never necessary—but just enough to have my feet on the ground before planning a great leap into the unknown. Perhaps I'll decide not to leap at all.

Next, I must have some idea of where to land at the end of my great leap. I must discover *what we would like things to be*. And before I can do that, I must discover who "we" are. Any real system involves many people, and when there are even two, there will be conflicting interests. And when there are conflicting interests, there will be political decisions, which may not be mine, as designer, to make.

When seeking his lady fair, the romantic soldier is captivated by appearance, rather than underlying character. Being Italian, he dreams of a tall blonde with blue eyes, and fails to notice the short, dark woman who would be a perfect soulmate. Perhaps it's not romantic to ask why he's enamored with height, but it might increase his chances of happiness. Once his desires have been specified functionally, he might discover that there are other solutions to his problem. For instance, athletic ability might correlate with height, and so might self-confidence. By specifying that he wants a self-confident, strong, and healthy lover, he might open his eyes to a woman a head shorter than he, but who happens to hold a gold medal in the 400 meters.

Starting from a list of functional criteria may not be romantic, but it enables me to use a third stage of design modeled on an evolutionary process. The list of criteria forms the environment to which the system I choose will have to be adapted. In my mind I run through dozens of variations of candidate systems, pruning away those that don't fit the environment and producing variants of those that fit best.

A recurrent romantic notion about design is that we will ultimately find the "best" design, if we use the right method. When I undertake to design something, my goals are far more modest. When I generate ideas, I'm not looking for the best, but only something that satisfies a reasonable set of functional criteria. When I start, I'm not certain that any system can meet my objectives, so why worry about some fantastic notion of "best"? I'm not even sure that I've got my objectives right, and in practice I'm all too eager to abandon some "essential" ingredient if a neat solution comes to mind that doesn't have it. If there is a good solution, I may not ever find it, so I might abandon the project or settle for less just to make life easier.

On the other hand, if solutions come too easily, I may simply raise my aspirations and make the job more difficult. In short, as a designer, I act more like a fickle young lover than a true romantic knight. People who teach design as if it were the quest for the Holy Grail or the faithful woman are doing a disservice to the richness that real-world design shares with such processes as biological evolution. They are also producing naive designers, ready to be duped the first time they fall in love with a real system.

Design must not be handicapped by a narrow vision that seeks to find *the* design, nor must it be handicapped by a narrow vision of "the real world." Success depends on the ability to generate ideas in sufficient variety to ensure that some will survive the screening process of reality. But success also depends on the ability to recognize when the screening is too harsh, and to relax some "real" condition. To generate ideas, I must have the romantic imagination of a young lover, but to control that imagination, I must also have some of the world-weary knowledge of a Don Alphonso. *The way things are* is always fighting with *the way things might be.*

Back and forth the battle rages—synthesis, analysis, then back to synthesis as the well of romantic imagination runs dry. And finally, the shape of a design emerges—not perfect in any respect, but not achieving false perfection through ignorance of reality.

This is my picture of design. As it stands, the picture is still a picture, too linear, too neat, too romantic for the reality of it all. I might start with a systems study—but I inevitably find that I really don't know the way things are until I try to change them. I might try to ferret out all potential objectives—but some new person or new idea always pops out of the woodwork just as I place the last piece of the design puzzle. Much as I wish that people wouldn't disturb my design, they all do it!

I believe that every real designer has some skeletal approach to design, but actually draws mostly on events and on personal experience with previous design problems. Few real designers carry a theory of design—even a rough one such as a mine—very far into the world of practice. Sometimes they'll drag out an old, world-weary design and paint its face with a rouge of theory—enough to fool a few romantic young designers, but not enough for us crusty old Don Alphonsos.

No, it's not our job to paint dreamy pictures of designs that cannot be realized in this all-too-real world. Still, the young romantic has one advantage. Without dreams, who will motivate the developers to build with sufficient romantic vigor? We need the grand visions of the young, or we perish.

That's the way we are! We're romantics because the "real world" is too heavy for us to bear alone. Without some romantic illusions, we would never fall in love, nor dare to design anything. A true designer has to be enough of a realist to know that any design is bound to be fickle, but enough of a romantic to try anyway. And a true teacher of design has to be tough enough to win a few bets, but romantic enough to let young love triumph in the end.

## How to Find Miracles

Would you like to see a miracle? Next time you're out walking, stand still by the side of the road, shut your eyes, count slowly to seventy-three, then open your eyes and notice the license number of the first licensed vehicle you see. Last time I tried that, I saw number 2-52783, clearly a miraculous number.

Why is 2-52783 miraculous? Well, around here the 2- isn't so special because all cars in this county have the 2- prefix. But 52783! Now there's something really special! Just think how unlikely it was to get precisely that particular number. Perhaps one chance in a million, or even less. That's as close to a miracle as you're ever likely to get.

John von Neumann is reported to have said, "Every license plate you see is a miracle." Von Neuman, even if he didn't actually say that, certainly understood the importance of miracles to computer programming. Programmers encounter miracles almost every day. On some days they encounter three or four. For instance, the CRIBBAGE programmer (see "A Cribbage Lesson") who thought a hand with four fives was "unlikely" was sure to be surprised by the number of "miracles" happening once his program became widely used—and failed every time it generated a hand with four fives.

Much of the time our impression of the miraculous arises from an insufficient appreciation of the ordinary. A cribbage hand of four fives seems miraculous in the light of our *personal* experience with cribbage. This experience differs from the computer's experience in at least two ways:

1. We play one hand at a time.
2. We play hands at a human rate, not an electronic rate.

In a lifetime, a cribbage player might play 100,000 hands, but a computer might play 100,000 hands in an hour's simulation. Or if a CRIBBAGE program is distributed to 1,000 locations, it might play 100,000 hands a day against human opponents.

What are the actual odds of getting four fives in one hand? Keeping the math simple, if we were dealt only four cards instead of six, the chances would be one in 270,725, so with six cards, our chances have to be better. Getting four fives, then, is the kind of "miracle" a cribbage player might experience once every couple of years of dedicated play. I personally can recall at least three occasions when I held four fives (and didn't cut a ten on any of them).

It's important to notice that I *recall* getting four fives, for that's part of what it takes to qualify a hand as a miracle. A miracle must be improbable, of course, but must also be noteworthy. That's why you weren't too impressed when I said that 2-52783 was a miracle.

Well, what would make a particular data case noteworthy to a computer programmer? The four fives was noteworthy because it caused the program to fail, not because it was such a great cribbage hand. A hand consisting of the three of hearts, seven of spades, eight of spades, and jack of diamonds isn't very noteworthy, but it would be if it was the one hand that made our CRIBBAGE program fail. And, from a probability standpoint, it's precisely as improbable as four fives.

From one point of view, each program that processes large amounts of data is constantly searching for miracles of this type. If there is any one case that makes it fail, that program will make us notice that case if and when it arises. If the program is widely used, then the chance of a miracle becomes even greater. No wonder programmers are familiar with this kind of miracle.

Some programmers like to be regarded as "miracle workers," but a more sensible approach is to define their job as "miracle prevention." That's the way the boss sees it, especially when the system fails on some "trivial" case. The boss sees the case as trivial because few bosses have been trained in the statistics of miracles. Most bosses, if they stood on the street and watched license numbers, wouldn't appreciate the glory and magnificence of license plates one little bit.

Training the boss to understand miracles may be difficult, but it's an essential step in programming success. There are some things the boss must be made to understand:

1.  Miracles can be reduced in number by sound programming practice.
2.  Miracles can never be eliminated by any practice, but merely reduced.
3.  As the number of miracles is reduced, the remaining miracles seem even more miraculous.
4.  If you start believing miracles can never happen, you start being extremely vulnerable to them when the inevitable is upon you. Then miracles become disasters.
5.  Some of the biggest strides in reducing the number and impact of miracles can be made *outside* of programming as it's normally considered.

Point 5 is essential. For instance, when you purchase software, you can reduce the number of miracles you experience by proper timing of your purchase. You needn't be the first on your block to get the new system. If you wait, other people will experience the joy of miraculous disasters, and you may be left with a rather dull but workable product.

The secret of timing your purchases depends on having a large population of users, each of whom might experience one or two miracles, rather than one user who experiences a thousand. The thousandth miracle no longer seems to be a heavenly act—if you've also had to live through the first 999. So if you're making or selling software, the same secret can be put to work to make your product seem less miraculous. Get the product spread around to as many users as possible, so each one experiences only a few "impossible" failures. That way they can retain their faith in your programming ability, which is, in fact, merely your ability to control the flow of miracles.

If you're working with a single user, you've got to take some other approach to miracle prevention. We've had great success involving people in contests to conjure up "pathological"—i.e., miraculous—test cases. Universities encourage students to attack security systems. Businesses can do it by offering small but symbolically significant prizes for thinking about the bizarre before it happens.

Here's one final example from outside computing. For many years now, I've been trying to find a rhyme in English for the word *orange*, but many people have told me it's impossible. I don't believe it's impossible, but I haven't yet been able to conjure up a counterexample. While discussing this problem recently in London, a Liverpudlian said that for many years he had been seeking a rhyme for the word *chimney*, and considered it impossible. "In that case," I said, "chimney 'rhymes' with orange, in the sense that they both belong to the class of things that have no other rhymes."

We had a lot of fun with that idea, and tried to think of other words in that class. Some candidates were *sugar* and *algebra*, both Arabic in origin and thus likely candidates for nonrhyming in English. I've been thinking about the question ever since, and have decided that a good approach to solving my problem would be to use my syndicated column as a way of broadcasting in the same way a software house broadcasts its problems when it distributes buggy products.

So here's my offer. I'm looking for a miracle—an English (or American) rhyme for *orange*. This is the test case that would make my "program" fail after all these years. I'll publish any miracles

I receive, and even some close tries that don't quite make it. Even better, of course, would be to embed the miracle in a proper poetic context, as I've done with the following *chimney* miracle I sent my friend from Liverpool:

When Santa played football, as kicker,
He developed one leg that was thicker.
He could start down a chimney
By inserting his slim knee,
But his fat knee was always the sticker.

As we Yanks always say, the difficult we do right now; the impossible takes a little longer.

# A Postscript on Miracles

True to prediction, the previous essay, when published in *Datalink*, produced a "miraculous" reply from Paul Coyle, who wrote as follows:

Concerning your search for a rhyme for 'orange,' it seems to me that your closing statement introduces yet another important variable which might even find your solution. I would class this as an 'environmental' variable. A package or system may run its usual routine way on one particular configuration, but change that configuration and miracles can happen. What's boring to Honeywell may be a miraculous interrupt and core-dump to IBM. The usual arguments about standard languages and compilers can also apply here.

"But what has that to do with oranges? Your closing statement was, 'As we Yanks always say . . .'; and there is your different environment. We (English/Yorkshire, etc.) tend to pronounce 'orange' with the 'a' having an 'i' sound—'oringe'—whereas you Yanks (your term!) tend to elide the 'r' into the 'n' with the 'a' almost disappearing— 'ornge.' So what rhymes with orange depends upon how you speak and who you are talking to—i.e. on your past and current environments. So forget your problem with the word itself and find somebody with the correct environment. Perhaps because of a dialect or speech impediment they fail to pronounce the opening 'o'—then 'range' gives you a perfect rhyme.

Because he's made me look so smart, I can't resist honoring Paul with a limerick:

An audacious young Yorker named Paul
Built a gingerbread door in his wall,
    With marmalade—orange—
    He greased up the door hinge,
But the bees made him pay for his gall.

The moral of the limerick, of course, is that not all clever ideas pay off. Still, we need all the ideas we can get, and the technique of broadcasting for miraculous ideas is a potent one for any designer to practice.

Why don't more designers use this powerful idea? Perhaps they're afraid to admit to the world that they can't come up with a wonderful, original solution to every possible problem. (Or is that "oranginal solution"?) If you're afraid to make a fool of yourself, you'll never amount to much as a designer. The ability to *play with ideas* is an essential characteristic of any successful designer or design team—if only because it helps them anticipate "miracles" in their system's environment.

**151**

A good test of your ability to play with ideas is your reaction to limericks. If you find them "wasteful," then perhaps you take yourself and your work too seriously. I don't mean you have to like *my* limericks, but, rather, that you've got to have the courage to say something foolish in public. Try writing your own limerick—as a test for your potential as a designer. A limerick has a lot of structure in its rhyme pattern, but that doesn't stop you from being so creative you could do something outlandish. The creative designer always takes a predetermined pattern and creates something outlandish within it, so the limerick can be your aptitude test.

Paul Coyle didn't send me a limerick, but his thought process is sufficiently audacious to convince me he'd make a good designer—at least on that account. Here's how he finished his letter:

> Dare I say with the multiplicity of dialects and accents (and environments) there may exist an individual who pronounces 'orange' as 'chimney,' although you might run into other communications problems. For instance, will the reverse apply and 'chimney' be pronounced as 'orange'? I suspect things aren't this simple. Try telling the fire brigade your orange is on fire!

Anyone who can play with ideas this way, and then shoot them down with the same ease, has a promising future. And although 'chimney' pronounced as 'orange' might not work too well, the idea of moving the system to a different environment is too powerful to ignore.

Daniel Freedman has a similar idea about mental illness. Just as pronunciations differ, so do cultures differ. What is crazy in one culture is perfectly acceptable—even laudable—in another. The answer to the mental illness problem is therefore a sort of travel agency that finds the right culture for each crazy person. I think the same sort of thing can be developed for computer people and their employers.

It also works for systems, as Paul suggests. I recall one large simulation model that never worked, and always crashed the operating system in peculiar and unpredictable ways. It was about to be written off as a total loss when the training department requisitioned it. They used it as a training simulator for machine operators. Put in the normal job stream, it would crash the system and allow the operators to practice their recovery procedures. Because it was so unpredictable, it gave them far more thorough training than any planned simulator could have done.

The moral is simple. If the shoe fits, wear it; but if it doesn't fit, perhaps you need a new foot.

# The Confusion Technique

Americans are great exporters of amusements—pinball machines, recordings, movies, and drugs. But one American amusement that has not reached Europe in any significant way is the drive-in movie. Is it the lack of the automobile culture and the greater cost of a giant parking lot that have kept the drive-in on this side of the Atlantic? Perhaps, but I have other ideas. The problem is that drive-ins, like pickles and sausages, are an acquired taste. You have to *learn* how to go to a drive-in movie.

Years ago the drive-in was merely a place to perform certain sexual rituals, safe from police and parental interference. Now that police and parents are busy exploring their own sexual potential, there are more convenient places for the youth of America to go for simple sex. Now people at drive-ins actually watch the films.

Yet the essential difference between a drive-in and a walk-in movie is still privacy. Although we may now be public about our sex lives, we must still keep our *opinions* to ourselves. In a walk-in movie, you may not laugh when others cry, cry when others eat popcorn, or talk (with or without popcorn) when others are listening. In the drive-in, however, you may carry on to your heart's content, as long as you have chosen your "car mate" with discretion.

Once you understand the essential function of the drive-in, you graduate to the question of selecting movies. Drive-ins always show at least two films at a time, and often three or more on weekends. Quite frequently, you start your menu with a film you have seen already downtown, but were unable to talk about. As second feature, you might choose a film you missed the first time around but wish you hadn't. But it's the third feature that's the prize. You always want at least one film on the program that will be so bad it will be good. From this past season, several examples come to mind, such as, *Orca the Killer Whale, Cheerleaders Sex Rally,* and *Super-Van Strikes Again.*

All of which leads up to the point of this essay. If something is sufficiently bad, its badness becomes a source of amusement or education. This circuitous introduction was not precipitated by a film, but by a book, *How Real Is Real?* by Paul Watzlawick. Now, *How Real Is Real?* is not really in a class with *Orca* or the *Cheerleaders,* and it will probably never be made into a film at all, let alone reach the drive-in, but it does have its moments. And those moments are instructive.

Watzlawick is a very intelligent person, full of nice bits of information which I intend to steal for other essays. The book—

subtitled *An Anecdotal Introduction to Communications Theory*—
attempts to develop the worthy theme that many problems may
be solved at the level of changed perception and communication.
And many of the anecdotes do demonstrate that point. But at bottom, the author is a "think tanker," and many of his examples are
as naive as scenes from *Super-Van*.

Watzlawick's most outrageous example was his opinion of
how the Patricia Hearst kidnapping ought to have been handled by
the investigating authorities. Watzlawick suggests they should have
used "Erickson's Confusion Technique," in the following way:

> Utilizing the same channels of delivery as the abductors, it would
> have been relatively simple for them to deliver to the mass media
> fake messages, contradicting the real ones but similarly threatening
> the life of Patricia Hearst if they were not complied with.

How does Watzlawick believe this would work? He goes on to
explain:

> Very quickly a situation of total confusion could have been set up.
> None of the threats and demands could have been believed, because
> every message would have been contradicted or confused by another,
> allegedly coming from the 'real' abductors . . .

Sound plausible? It's about plausible enough to escape scrutiny
downtown, but it would never play at the drive-in.

Oh, Watzlawick—in true think-tanker fashion—sets up some
straw men in opposition to his pet idea, then knocks them down,
as when he writes:

> Needless to say, in our era of alarming electronic progress, the pro-
> duction of perfectly genuine-sounding tapes would have presented
> no technical difficulties whatsoever.

Let's ignore the ignorance this displays of the techniques for de-
tecting counterfeit tapes. Notice instead how enthusiasm for one's
pet idea can cloud the mind.

It's very difficult for me to believe that Watzlawick ever
thought critically about this idea for fifteen seconds, but its naiveté
is typical for this genre of speculative systems writing. Think for
a moment with the mind of the field marshal of the Symbionese
Liberation Army. Would you allow the confusion technique to get
in the way of your world-shaking plan? Not on your life!

People in ivory towers can think very effectively in the idiom of their clients—government leaders, corporate executives, and the like—but for that very reason, perhaps, they can't bring themselves to think like the mass of people who don't sit in high positions of power. I'm not being critical, for each of us is severely limited by the blinders of our culture, our education, our job, and our social circle. That's why we don't all laugh at *Orca the Killer Whale.* That's why nobody can be allowed to make serious systems design decisions without submitting those decisions to most rigorous and diverse tests.

Sometimes we can test the decisions on the very people who will be affected by them, but in many life-and-death situations, that kind of testing won't be possible. In those cases, the ivory-tower designer has got to attack every idea and assumption with the same vigor we usually reserve for drive-in criticism.

Suppose, for instance, we were writing a script for a kidnap movie. The heiress is kidnapped and the investigating authorities put the confusion technique into action. Then Field Marshal Cinque, not being constrained by the niceties of the upper classes, simply authenticates his next message by sending along one of his captive's fingers!

But that's revolting, you say. Unthinkable! Yes, it's not very nice, but who said terrorists were nice? Coming from another world than the systems designers, they can be expected to think differently. And, in this case, more effectively—with or without doctorates. Even with "alarming electronic progress," we're not likely to be able to fake a fingerprint if it's accompanied by the finger. And by the time the field marshal runs out of message verifiers, the victim's family will have long since run out of patience with the investigating authorities and their oh-so-clever advisors.

Unhappily, this is not a far-fetched example. I run into similar modes of thinking every time I examine grand systems designs. For instance, the software and hardware experts design the "impenetrable" curbside cash dispenser—only to have the crooks drive up in a van and remove the entire dispenser using jackhammers. There is simply too much distance between the high-level designers and the people with whom their systems are supposed to work. This applies to security and privacy systems as well as antiterrorist systems, but it also applies even more strongly to the most mundane data processing systems we can imagine. As systems get more vast, more complex, our techniques of thinking about them have to get tougher and more realistic. Of course we need high-level abstraction, but not at the price of losing touch with day-to-day reality.

And where does that reality start? I'd say the drive-in movie would be a pretty good place. That is, let's put all those great ideas through a second run and let everyone laugh at them in the privacy of their family automobiles.

# WIGGLE Charts—A Sketching Tool for Designers

*There's no sense being precise about something when you don't even know what you're talking about.*

John von Neumann

For systems designers, it is the best of times and the worst of times. For years we muddled through with a few simple graphic tools for design and documentation—flowcharts, block diagrams, and perhaps decision tables. Then came the diagram explosion, with HIPO, HIPO/DB, Warnier-Orr diagrams, Softech's SADT, Nassi-Shneiderman charts, Petri nets, Constantine structure charts and data flow diagrams, Jackson data structure diagrams, and coding schemes. And for each of these diagrams, you need only bend a line or add a symbol to get your name in the papers as the inventor of yet another graphic design tool.

Although the choice is large, it is really not very wide. Each of these diagrammatic schemes shares the characteristic of *precision*—wonderful when you know what you're talking about, but time-consuming and thought-stifling when you don't. And, since most design work is spent thinking *roughly*, few of these diagrams are of much help through large parts of the design process.

In other design fields, such as architecture, the rough sketch is the most frequently used graphic device, and precise detailed drawings are rarely used at all until the *creative* part of the design work is finished. The rough sketch has several advantages over the precise drawing:

1. It can be drawn much faster, thus using less time.
2. It represents less investment of time, so we're not afraid to throw it away and try something else.
3. It's very roughness conveys important information about where we are in the design process.

In data processing, rough sketches have always existed, but have never been glorified by a name or by favorable publicity. Schools of architecture offer courses in sketching. The student architect who makes clear quick sketches is much admired by faculty and peers alike. It's time we learned from more mature disciplines and put sketching up on a pedestal.

For many years, I've taught a method of sketching that can be used with most of the diagrammatic techniques now used in data processing. Although it's been received with enthusiasm, it's never received much publicity, perhaps because:

1. It doesn't require a template.
2. It doesn't have a name.

Although I'll continue to resist the template forces, I've decided to bring the baby to life with a catchy acronym, WIGGLE Charts, for Weinberg's Ideogram for Generating Graphics that Lack Exactitude.

A WIGGLE is merely a box, or block, or line, with one or more rough edges. The rough edges indicate what parts represented by the box or line are imprecisely known. For instance, Figure 17 is a sketch of a system using a block diagram form. Each box represents input coming from the left, processing inside, and output going to the right. Box 1 has a straight line at its left side, indicating that the input to Box 1 is clearly defined somewhere. The right side, however, is rough, indicating that we haven't decided what its output will be. All we know, as indicated in the diagram, is that *some* output will be passed to a second box. The top and bottom of Box 1 are rough lines, indicating that we don't know exactly what this process will be.

Box 2 has undefined input and output, but its process is well known to us, and clearly delimited in scope. Perhaps we have decided to use a package sort, though we don't know which one, so we haven't decided upon a record format.

Box 3 takes the unknown output of Box 2 as its unknown input. By a process that's not yet well defined, it produces two outputs, one well defined and one known only roughly. Perhaps the first report is defined by legal requirements, or by input needs of another system, while the second output is an error report whose format is left open at this stage of the design process. The rough

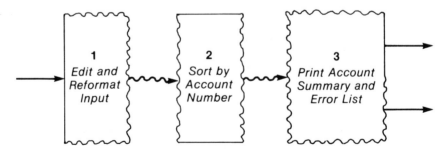

**Figure 17.**   A WIGGLE Block Diagram

arrows between the boxes indicate that we haven't yet decided how control will pass from one box to another. They could be subroutines of the same master routine, or steps in the same job, or separate steps manually coordinated.

Taken together, these three WIGGLE boxes and their arrows give a sketch of the overall design we have in mind. Perhaps more important is what they *don't* do:

1.  They don't give us or any reader an unjustified feeling of precision.
2.  They don't intimidate anyone who has an idea about changing something that might improve the design.
3.  They haven't wasted a lot of time drawing with templates.

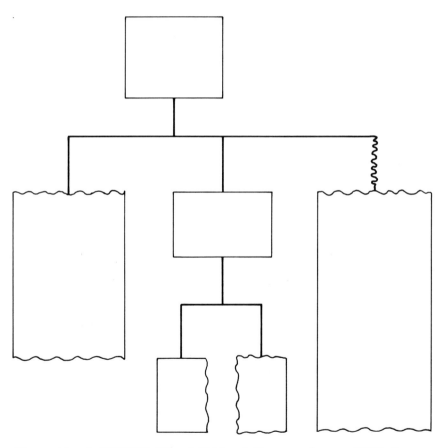

**Figure 18.** A WIGGLE Visual Table of Contents from a HIPO System (Each box has input on left end, output on right.)

Perhaps the nicest feature of WIGGLE charts is that they can be used with just about anybody's diagrammatic technique. Figure 18 shows a Hierarchic WIGGLE, or a WIGGLE Visual Table of Contents (WVTOC) for use with a HIPO system. In this application of the WIGGLE, as in Figure 17, the overall *size* of the boxes can be used to indicate (roughly) how big an effort we see in building this box. Alternatively, it can be used to approximate how much execution time or other resource we expect to be consumed here.

Figure 19 shows a Nassi-Shneiderman WIGGLE. In this chart, the size of the wiggles in the diagram indicates roughly *how* uncertain we are of the particular part of the design. The vertical loop wiggle is quite small, perhaps indicating that we're not sure if the loop is to be done $N$ or $N+1$ times. Similarly, the slanted wiggles

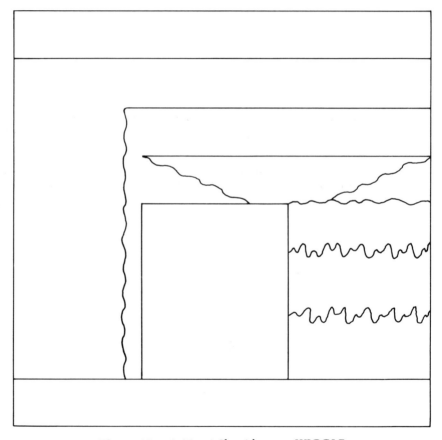

**Figure 19.** A Nassi-Shneiderman WIGGLE

on the decision are small, indicating perhaps that we don't yet know just where the "equal" case will go. But the large wiggles dividing the right branch of the decision into three boxes are very large, indicating great uncertainty about the functions to be performed here.

Each of these charts should be sufficiently "clear"—as sketches—to readers who can read the original, unwiggled, chart. Just for completeness, Figure 20 shows a key to the use of the WIGGLE. Using this figure, you should be able to begin sketching your favorite design pictures using the WIGGLE, even if you're still in love with good old flowcharts.

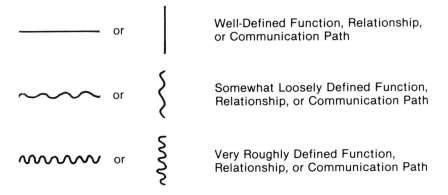

All these conventions may be applied to sides of boxes, regardless of the shape of the box, as well as to arrows or other lines connecting boxes.

**Figure 20.** The WIGGLE System Condensed to a Few Simple Rules That Can Be Applied to Any Graphic Scheme

## Featuring Failure

It's a bit disconcerting to watch the younger generation attacking their personal computers with the same naive vigor we had when we programmed our first machines, oh, so many years ago. It's tempting to say things like, "Haven't they learned *anything?*" but the real annoyance is that they're learning *everything*. All of our hard-won secrets are now open to anyone with the price of a PET, TRS80, or APPLE. In our day, the apple was for the teacher, but now the APPLE is the teacher. What's the advantage of growing old if you can't tell the young quaint stories about the old days?

Even worse, of course, is when these young upstarts think they can teach us something we don't already know. A few weeks ago, one young enthusiast was showing me his personal text editor, bragging about its many features crammed into a small memory. I was taking it pretty well until he explained to me that one of the best features was the editor's ability to recognize commands in abbreviated forms. "In fact," he boasted, "they'll recognize almost any sensible abbreviation, or spelling error, like CHG for CHANGE, or JFY for JUSTIFY."

He wasn't going to catch me there. I'd heard that story too many times before. "How about GUSTIFY?" I asked, innocently.

"That's not a sensible abbreviation."

"It's a plausible spelling error."

"I don't think anyone would misspell justify with a g."

"I just did."

"Well, you're being unreasonable."

Eventually, I got him to admit what I knew all along. His "spelling correction and abbreviation" algorithm consisted in taking the first letter of the command and ignoring the rest. Given the small amount of main memory, he could hardly have done anything else. In fact, he would have been in trouble if he had to recognize even the full correct names of his twenty-one commands. To store the names for comparison would have taken at least 100 bytes, and probably more than 200.

What this tyro had done was rediscover one of the oldest problem avoidance tricks in the programmer's book. As a designer for a large computer manufacturer put it, "If you can't fix it, feature it." In other words, if we can't provide what the customer wants, then convince the customer that what we do provide is what was really wanted.

If this sounds a bit cynical, it's only because you've not explored the way other professionals do business. Consider the doctor who can't cure your arthritis, but becomes a hero for writing a prescription for a powerful pain killer—which you wouldn't have needed if you'd been truly cured.

Many of today's popular drugs are used not for their original purpose, but for an effect that was considered a "side effect"—often undesirable—in the beginning. Birth control pills are one example—sterility was a side effect in the original use, now forgotten.

A neighbor of ours had the misfortune to strike a gigantic rock when excavating for an addition to his home. It would have cost a fortune to remove, so the architect replanned the room around this "natural stone wall."

Some years ago, in New York, a real estate saleswoman showed us a house next to a sawmill. It was a great location, she told us, because it was so loud during the day that nobody else lived in the neighborhood. We worked during the day, most of the time, and wouldn't hear the saw except early in the morning, which would save us having to set an alarm clock.

Much of the dissatisfaction with computers arises from unkept promises. Sometimes the programmers can't do what the specification has promised. More often, perhaps, they get bored with the entire project and don't want to add what they consider "frills." We could avoid most of this trouble by holding our tongues at the outset, when we're inclined to make promises we'll regret. If we leave the specification sufficiently vague, we'll have more room to maneuver once we discover what we can't or won't do. If we do it right, we can even convince the client that the failures are features.

We've performed or watched a number of experiments with developing information systems in stages. We start with a skeletal system that does some minimal job. After the client has had a chance to get the feel of the new system, we sit down and specify the next round of incremental improvements.

In our early experiments, we'd try to specify all six or eight stages at the outset—design in advance but build incrementally. Almost every time we reached one of the later stages, the client would either change the specification drastically or simply indicate that the last stage wasn't necessary. Our first reaction was that we had failed to give the clients what they wanted. Later, we realized that the ability to decide when to stop was a *feature*, not a failure.

After that, we refrained from specifying too many stages in advance, except in the broadest sketches. With so little invested in specifications and design, neither we nor the client felt robbed if we decided to change some original idea.

A famous American architect once said, "An architect's first job is to take what's given and change it." I would go one step further to say, "A program developer's *continuing* job is to take what's given and change it." I'm not saying that the trend toward carefully written specifications is wrong. I am saying that *over-specifying* and *premature specifying* are wrong. Don't allow yourself to be pinned down! Not unless it's absolutely necessary. Avoid problems by getting sufficient agreement, in advance, to know that the client will be satisfied in the major things. Avoid eventual problems by not allowing the client to fence you in with minor things that you'll be unhappy doing later on.

When the job of the systems analyst is separated from the job of the programmer, the tendency is to overspecify from the outset. The analyst wants to finish a job quickly, and not have to stick around for the grisly details. On small, familiar jobs, that pattern works well. As jobs grow larger, or more experimental, it's the road to sorrow. Toward the end of an overspecified project, the programmers are so bored and frustrated that they start leaving like lemmings. Everything was laid out in advance, so they have no room to be heroes by providing features. The best they can do is what the analysts had written down two years ago—not very exciting. The worst they can do is take some shortcut to get the grungy mess over with a bit sooner, even if the shortcut is something the client would dearly love.

I'll never forget the reaction of the first client who learned that he could terminate his project when only about eighty percent of the specification had been implemented. "Do you mean it's finished?" he asked, in disbelief.

"Yes, finished."

"You mean I can just keep on using the system as it is, and don't have to suffer through any more updates?"

"That's right—unless you decide you want something later on."

"I suppose I'll have to pay the full estimate anyway?"

"No, just the amount that's been spent so far—about sixty percent of the original figure."

"Do you mean that the last twenty percent was going to cost me forty percent of my money?"

"Only if we made the budget. More likely it would have cost you half or more, if we'd overrun."

"But I'm done three months ahead of schedule. I would have been willing to pay extra for that."

"That's all right. It's a bonus because of our superior methods."

I know it's hard to believe if you haven't seen it happen, but I've seen it more than a dozen times and heard about others. It does sound a bit like the proposal to eliminate all the auto collisions by eliminating half the autos—since two cars are involved in each collision.

In our case, we could start with the observation that ninety percent of the aggravation in programming projects comes at the end. Therefore, the way to get rid of most of the aggravation is to chop off the end—the part where you're getting every last bit and twiddle in place. That would be ridiculous if the most important matters were left to the end, so make your job easier by doing the important things first and never doing the unimportant ones. Or hardly ever.

## A Rose and a Rose: A Fable

One year, at the American Beauty Rose Finals, the judges were completely split between the lovely pink Rose Arizona and the blushing red Rose Washington. While the crowd waited breathlessly for the announcement of a decision, the two contestants sat together off stage and discussed their prospects.

"I'm sure that you will win," Rose Washington was saying. "Your pale loveliness can't help but make the judges pick you. I've tried to achieve that coloring, but in Washington it rains too much and spoils my petals."

"Oh, no," Rose Arizona contradicted, "Your deep, delicious red is the color every true rose should be. In Arizona, though, there is too much sun and bright colors fade away into dull pink like me."

"I know that you're wrong," replied Rose Washington, "but if I did win, I know one thing—I'd spend my year's reign in a hothouse where it never rained at all, and the sunlamps were turned on even during the night."

"If I were lucky enough to win, I certainly wouldn't do that. I'd request a hothouse where the glass was covered and the sprinklers were turned on with nice, cool water twenty-four hours a day."

In the meantime, the judges had agreed that they couldn't choose, so they announced that there would be two American Beauty Rose winners that year. Pink Rose Arizona and red Rose Washington each shed a dew-drop tear as they began their year's reign. They were so busy during that year that they didn't meet again until they came the next year to Atlantic City to present prizes to the new winner. Each was shocked at the changes in the other.

"Oh, dear Rose Washington," gasped Rose Arizona. "What has happened to your lovely red petals?"

"I'm afraid that all that sun has utterly ruined my complexion. Look how dry and crackly my petals are. Why, they are so brittle that I'm afraid to be sniffed for fear of having them knocked off by somebody's nose. Oh, how I long for the moisture of my dear state of Washington! But you, my friend, haven't you had moisture all year long? What has become of you?"

"As you can see, my pink has turned to gray, not red, and I'm so waterlogged that I can't curl my petals properly. At least I don't have to worry about being sniffed. In one year without the sun, I've lost my perfume completely."

And so, after presenting Rose Wisconsin, the new champion, with many prizes and a bit of advice, the two friends parted for their homes. Soon, they each recovered most—though not quite all—of their former beauty. Though they were perhaps not as beautiful as before, they were completely content with what they had.

MORAL:   *Beauty requires both sun and rain—and is more rooted in familiar surroundings than we'd like to think.*

IN OTHER WORDS: The most beautiful design is always achieved with some sort of balance—and sometimes that balance is best achieved by leaving the system untouched in the environment to which it has become adapted.

# PART VIII

# Epilogue

The essay you are about to read was not in this book's original manuscript. When Chuck Durang, my editor, read that manuscript, he thought it was terrific—except "that it sort of ends with a whimper, not a bang." My first reaction was to recall the gag line, "It's a great symphony, Ludwig—except for the first four notes!"

Next I plunged into my poetry library to find the context of the original line, possibly as a source of inspiration. That activity managed to consume a few hours and remind me of the great fascination poets have with death and the end of the world. Eventually I got a grip on myself and said, "Well, Jerry, you've had an epilogue on most of your books. What is it you're trying to avoid here?"

Looking at the last fable in this volume, I can see that I was trying to whimper a message that I was afraid to bang. So let me see if I have the courage to state it a bit more forcefully.

1. Systems analysis and design have solved many problems, but they have also created many problems.
2. Some of these problems have been created by *misapplication* of good analysis and design principles.
3. Some of these problems have been created by *ignorance* of good analysis and design principles.
4. But there are many problems caused by intelligent and informed application of good analysis and design principles, problems that will not go away no matter how cleverly we rethink systems analysis and design.
5. Those problems arise when we are applying systems analysis and design in places that were best left alone.

Now that I've stated that forcefully, let me hammer it home with a few examples. As a writer about information systems, I have a vested interest in change. If things were not changing so fast, few books would be needed. If few books were needed, few authors could make a comfortable living. I'm not rich, but I like the financial independence I've achieved by writing about how new ways of thinking can improve our lives. If these approaches were to fall into wide disfavor, my royalties would decline. Other people have similar vested interests in change—just as others have vested interests in stability. If technological change fell into disfavor, systems analysts and designers would soon be drawing unemployment compensation.

So it's easy for some of us to believe that change is always good. That's why we ought at least to think of the possibility that some system we are attacking might be better off just now without our services.

Here's another example of inappropriate systems work. Most of us have had the experience of filling out tax returns. As systems analysts, many of us have shared the thought, "Why does this have to be so complicated?" To answer such a question, you could study the evolution of the tax laws. No doubt you would find tales of special interest, but you would also find a blizzard of words on the subject of fairness to the "ordinary" taxpayer. Eventually you would see that most of the complication in tax laws arises from the constant compromising between these and other political forces.

But there's another level at which the question could be answered. The more complicated a tax system becomes, the more difficult it is for the "ordinary" taxpayer to use it effectively. If you are paying $100,00 in taxes, you can probably afford to hire a rather clever tax analyst. But if you're paying only $1,000, you can barely afford to pay H & R Block to get the forms right.

Other things being equal, complexity favors the rich. As a systems analyst, you can be used just as the tax analyst is used. Nothing wrong with that, if that's the way you want to be used. But if you imagine that you're favoring the poor, the tired, the huddled masses, you may want to examine what you're doing.

Such reversals of the apparent thrust of an effort are well known to systems analysts, but we seldom look for them in our very own work. Philosophically, you may favor the rich over the poor, but your work may not actually be supporting the rich even though it may seem so at the first level. You may be working as a systems analyst for a large established corporation, but change doesn't always favor the establishment—even when that's the intention.

In short, if you take more of a systems view of your own work, you may find that your professional life is not in consonance with your political and philosophical views, whatever they may be. Under those circumstances, doing a good systems job may make you most unhappy. The fable of the roses also spoke about balance—

a vague word, a whimpering word. I think now that what I meant by "balance" was *protecting the world from systems people by leaving a little margin for our own errors or misguided motives.*

We naturally feel we know much more than our predecessors, and we're probably right about that. But where do we really stand relative to all we might know? What do we really know about the systems we study? And what do we really know about *ourselves,* and why we study systems in the first place?

I have a theory that writers and systems analysts alike are responding to one of the basic human needs. Psychologists tell us about air, water, food, and sex, but there's another need that should be on the same list. This is the need to judge other people. Everyone has this need, though they may not be proud to admit it. You can survive a few weeks without food, but when was the last time you went two weeks without passing judgment on someone else's behavior?

The need to judge is much like the need for sex. It's not the kind of need we like to confess in public, but in private we love to satisfy it. Most people can keep their sexual needs under control when they threaten to be too disruptive of their social life. A few cannot, like the hapless creatures we call exhibitionists. But other people take the same needs and convert them into a marketable commodity. Instead of being labeled exhibitionists, they are called body builders or movie stars, and we all love them.

Writers, like me, are people who have similarly converted their need to judge into a marketable commodity. To write a book about other peoples' behavior, you have to have an uncontrollable urge to snoop and pass judgment. You also have to have an incredibly overinflated view of yourself.

It takes the same uncontrollable needs and inflated self-image to be a systems analyst/designer—to study what people do and tell them how to redesign their activities. In other contexts, at other times, we would have been called moralists, or do-gooders. And we all know how much good the do-gooders have done!

In view of all this, I can't resist one more piece of moralizing. I'm just sure the world would be a better place if writers and systems people would pause now and then to remind themselves of how little they really know. If they did, they would certainly allow for more "balance" in their words and in their works.

How can you achieve this kind of balance in your thinking? For one thing, you can establish the habit of asking yourself such questions as these:

1. Am I fostering change for the sake of change?
2. Am I serving interests I'd rather not serve?
3. Am I responding to inner impulses I'd rather not admit to having?
4. What are the chances that I may simply be wrong?

Paradoxically, the best way to avoid being a know-it-all is to keep learning more and more. You might think that your brain will eventually become so heavy that your head will fall off, but it won't. Indeed, the more you put in your head, the smaller it gets.

Well, Chuck, those are a few of the things I was trying to say with a whimper. I'm not sure that banging them out directly will have any more effect than coding them into parables. The Bible has survived a long time doing both, but there are still one or two people who commit adultery. Taking your advice, I've tried both paths, but somehow I doubt I'll reach ever a biblical level of efficacy. I may have an inflated self-image, but not that inflated. If you want any more bang, you'll have to supply each reader with a great big stick for applying to the forehead, or to the other end.

# Appendix: Description of the Black Box System

The black box as seen by the students is best described by showing the assignment as given to them (Figure A) and a sample of the output from one of the subsystems (Figure B). (The reader may wish to try "solving" the system shown in Figure B before proceeding,

### General Systems Theory
### Special Assignment: Studying the Black Box

Each team will be given a program which represents a computer model of eight (8) systems. The team is to conduct input-output studies of these systems to find out as much as possible about them without studying their internal construction. It is against the rules, therefore, to take dumps or to use any other information about the internal structure of the systems.

The only permitted method of analysis is to try different inputs and observe the outputs they produce. An input to one of the systems is constructed by entering a single line in the following format:

| Field # | Description |
|---------|-------------|
| 0 | System number (1–8) |
| 1 | Input variable number 1 |
| 2 | Input variable number 2 |
| 3 | Input variable number 3 |
| 4 | Input variable number 4 |
| 5 | Input variable number 5 |
| 6 | Input variable number 6 |
| 7 | Input variable number 7 |

All fields are integers and must be less than 1000 in absolute value, except field 1 which is limited to two digits. Only one number may appear in any field, and no field should be totally blank.

An identification line should be before the input lines, with the team number and the team password.

The blackbox will not accept input to a subsystem which has been used 100 times by the team. A report on the status of the team file will appear at the end of each run. You must, therefore, plan your studies carefully so you do not exhaust your 100 observations per subsystem too quickly.

After a reasonable amount of work, the team is to prepare a written report on what it discovered, how it discovered it, and what was learned about systems in the process.

**Figure A.** Class Assignment for the Black Box Exercise

| The Subsystem used is 4 | | | | | | | | | | | | | |
|---|---|---|---|---|---|---|---|---|---|---|---|---|---|
| Input Vector | | | | | | | Output Vector | | | | | | |
| 1 | 2 | 3 | 4 | 5 | 6 | 7 | 1 | 2 | 3 | 4 | 5 | 6 | 7 |
| 1 | 2 | 3 | 4 | 5 | 6 | 7 | 2 | 3 | 4 | 5 | 6 | 7 | 1 |
| 2 | 3 | 4 | 5 | 6 | 7 | 1 | 4 | 5 | 6 | 7 | 1 | 2 | 3 |
| 4 | 5 | 6 | 7 | 1 | 2 | 3 | 1 | 2 | 3 | 4 | 5 | 6 | 7 |
| 1 | 2 | 3 | 4 | 5 | 6 | 7 | 2 | 3 | 4 | 5 | 6 | 7 | 1 |
| 2 | 3 | 4 | 5 | 6 | 7 | 1 | 4 | 5 | 6 | 7 | 1 | 2 | 3 |
| 4 | 5 | 6 | 7 | 1 | 2 | 3 | 1 | 2 | 3 | 4 | 5 | 6 | 7 |
| 1 | 2 | 3 | 4 | 5 | 6 | 7 | 2 | 3 | 4 | 5 | 6 | 7 | 1 |
| 2 | 3 | 4 | 5 | 6 | 7 | 1 | 4 | 5 | 6 | 7 | 1 | 2 | 3 |
| 4 | 5 | 6 | 7 | 1 | 2 | 3 | 1 | 2 | 3 | 4 | 5 | 6 | 7 |
| 1 | 2 | 3 | 4 | 5 | 6 | 7 | 2 | 3 | 4 | 5 | 6 | 7 | 1 |
| 2 | 3 | 4 | 5 | 6 | 7 | 1 | 4 | 5 | 6 | 7 | 1 | 2 | 3 |
| 4 | 5 | 6 | 7 | 1 | 2 | 3 | 1 | 2 | 3 | 4 | 5 | 6 | 7 |
| 1 | 2 | 3 | 4 | 5 | 6 | 7 | 2 | 3 | 4 | 5 | 6 | 7 | 1 |
| 2 | 3 | 4 | 5 | 6 | 7 | 1 | 4 | 5 | 6 | 7 | 1 | 2 | 3 |
| 4 | 5 | 6 | 7 | 1 | 2 | 3 | 1 | 2 | 3 | 4 | 5 | 6 | 7 |
| 1 | 2 | 3 | 4 | 5 | 6 | 7 | 2 | 3 | 4 | 5 | 6 | 7 | 1 |
| 2 | 3 | 4 | 5 | 6 | 7 | 1 | 4 | 5 | 6 | 7 | 1 | 2 | 3 |
| 4 | 5 | 6 | 7 | 1 | 2 | 3 | 1 | 2 | 3 | 4 | 5 | 6 | 7 |
| 1 | 2 | 3 | 4 | 5 | 6 | 7 | 2 | 3 | 4 | 5 | 6 | 7 | 1 |

**Figure B.** A Typical Set of Student Inputs to the Black Box, along with the Outputs They Received

in order to get some idea of what the students are up against.) On the inside, of course, the system consists of a set of subprograms under the control of a master program that selects the appropriate one according to the first input value. The following is a description of eight of the systems we have used, along with an explanation of the principles they are intended to teach. In the descriptions, $V$ represents the input vector ($V_i$ is the $i$th element), $W$ represents the output vector, and $R$ is a matrix of constants that are fixed for any one team.

Figure A gives a summary description of each box, and Figure B gives a summary of the results obtained, in terms of the percent of students reaching a given level of "solution."

## Subsystem 1

$$W = aV + b$$

In this simple system, each element undergoes the same fixed transformation. The system is decomposable into seven independent subsystems, and is intended to show how decomposability leads to simplicity of observation. Subsystem 1 is indeed the simplest for the students to solve, on the average, and serves as a rather easy introduction to the mechanics of the black box.

## Subsystem 2

$$W = RV$$

This matrix-by-vector multiplication yields a simple linear transformation, which, if they have learned their lessons about linearity, the students should be able to identify rather rapidly. In solving this system, students get a pragmatic demonstration of how tests for linearity may be applied and how linearity simplifies the task of analysis. The special role of the unit vectors and the zero vectors is reinforced by studying this system.

## Subsystem 3

$W = RV$, if the sum of the squares of the elements in $V$ is less than some cutoff value. If the sum of squares is greater than the cutoff, one is added to the first element of $V$ and then the linear transformation is applied.

It is intended that the students will quickly recognize the relationship between this system and the second system. If the cutoff value is sufficiently high, some of the students will fail to notice that there is any difference at all between the two systems. When this flaw in their observation technique is revealed to them, they will have learned a lasting lesson about the necessity for making a few observations near the possible extremes. Those who do not notice the similarity to system 2 at all, of course, learn a valuable lesson in relating knowledge already gained to the problem at hand.

Primarily, however, this subsystem teaches about the difficulty we sometimes encounter in practice of defining a boundary precisely—even when a precise boundary actually exists. Each of the student teams that discovers the piecewise nature of this system forms some operational rule for distinguishing which transformation will be applied, but it is a rare team that induces it exactly. Such rules as "whenever there is an element greater than 10," or "whenever the sum of the elements is greater than 45," are quite frequent. Attempts to find a more precise boundary often lead to

| System | Transformation |
|--------|----------------|
| 1 | $W = aV + b$ |
| 2 | $W = RV$ |
| 3 | $W = RV$ or $W = R \cdot V'$ if $\Sigma V^2 > c$ |
| 4 | $xi = V_i \bmod 7 + 1$ <br> $W_i = V_{xi}$ |
| 5 | $W_i = S = \sum_{i=1}^{6}(-1)^i v_i v_{i+1}$ |
| 6 | $W_i = (n \bmod p)^2 / i$ <br> $n$ = card number <br> $p$ = page number |
| 7 | $W(t) = k(V(t) + W(t - \Delta t))$ <br> where $\Delta t = 1$ or 2 |
| 8 | Almost random pattern |

**Figure C.** Summary of Box Contents

| Box | Concept | Success |
|-----|---------|---------|
| 1 | Decomposible linear transformation | 100 |
| 2 | Linear transformation | 94 |
| 3a | Linear transformation | 98 |
| b | Existence boundary | 83 |
| c | Exact nature of boundary | 17 |
| 4a | Existence of control variables | 89 |
| b | Exact nature of control | 63 |
| 5a | Symmetry | 92 |
| b | More than symmetry but not complete | 51 |
| c | Complete solution | 23 |
| 6a | Some pattern | 77 |
| b | Complete pattern | 12 |
| c | Give up as partially "random" | 67 |
| 7a | Complete solutions: delay = 1 | 31 |
| b | Complete solutions: delay = 2 | 0 |
| 8a | Some pattern | 81 |
| b | Use statistical tests | 24 |
| c | Give up as partially "random" | 78 |
| d | Some nonexistent pattern | 58 |

**Figure D.** Various Kinds of Success in Extracting Laws

rules of great logical complexity, each of which fits the test cases that team tried, but each of which fails if tested on the cases of some other team. Here, then, is also taught the lesson of the limitations of induction, and of the necessity for choosing cases that try to *disprove* one's hypothesis, rather than prove it. Here also is taught the pervasive nature of our ideas of the linearity of the world, which makes the sum of squares a particularly difficult boundary to discover.

A few of the teams manage to ignore the problems of this system by dismissing the deviations about the cutoff as "not significant." Since they have been given no criterion for significance, this behavior may be used to illustrate how we often define what interests us by what we can easily understand.

## Subsystem 4

In this subsystem, the input variables play quite a different role than they did in the previous systems. In systems 1, 2 and 3, the input variables were, in a sense, "action" variables being transformed by the system. In this system, however, they are both "action" and "control" variables, in that their values are used as output, but the order of values is permuted depending on the values themselves. Thus, to get the value of the $i$th output element, the $i$th input element is taken modulo 7, one is added to the result to get it in the range 1 to 7, and that result is used as a subscript to select which input will be output. For instance, if the input vector is

$$1, 1, 5, 5, 9, 9, 15$$

the output will be calculated as follows:

$$w_1 = w_2 = v_2 = 1, \text{ because } 2 = 1 \bmod 7 + 1$$
$$w_3 = w_4 = v_6 = 9, \text{ because } 6 = 5 \bmod 7 + 1$$
$$w_5 = w_6 = v_3 = 5, \text{ because } 3 = 9 \bmod 7 + 1$$
$$w_7 = v_2 = 1, \text{ because } 2 = 15 \bmod 7 + 1.$$

Thus, the output vector is

$$1, 1, 9, 9, 5, 5, 1$$

In a sense, this transformation is a switching network, where the items switched are the inputs and the setting of the switches is also determined by the inputs. This black box can prove quite difficult for some of the students, demonstrating to them the different conceptual roles of action and control variables. Some of the difficulty also comes because the same variables are used in both functions, and it is thus hard to keep them straight in the conceptual model. Inasmuch as variables in real situations almost always have both functions intermixed, this lesson is a valuable one.

## Subsystem 5

This subsystem sets all of the output components to the same value. This value is determined by the value of the sum

$$S = \sum_{i=1}^{6} (-1)^i v_i v_{i+1}$$

One of the first lessons this subsystem teaches is how a many-one mapping reduces the information available in the input. In a sense, this function is a classifying function, which groups discriminably different inputs into equivalence classes. Such a classification may be related to a concept, though we may not be able to put a name on that concept. In this case, one part of that concept is symmetry, for any input which is symmetric left-to-right and right-to-left will yield a zero value for S. Other inputs, however, such as (0, 1, 0, 0, 0, 0, 1), which are not symmetric, will also yield zero, so that more than one nameable concept is lumped together.

Very few of the students solve this system completely, but nobody fails to find out something about it. It teaches that we can often gain incomplete but precise information about a system (as well as complete, but imprecise, information, as in subsystem 3).

The existence of symmetry as a part of the determination of the transformation teaches the students about the importance of choosing the appropriate level of observation of the system. Lower level concepts, based on the individual numbers, are of no help whatsoever in solving this system.

## Subsystem 6

In this system, the output value does not depend in any way on the supposed input (V), but depends instead on two partly concealed inputs, namely, the page number, $p$, and the number of the card or line in the input, $n$. The formula for the $i$th element of the output is

$$w_i = (n \text{ modulo } p)^2 / i$$

so that, for instance, if the 46th line appears on page 6, $w$ will be

$$(16, 8, 5, 4, 3, 2, 2)$$

There is much obvious pattern to the output of this system, but the students are rather quick to say that there is some "random" element involved. One of the lessons it teaches, then, is that what look like inputs are not always inputs. Furthermore, what look like inconsequential outputs (page number) are not always inconsequential. Here, then, the students learn that widening the vision in another way—outside the given description of the problem—can be just as important as achieving the proper level of observation (as in subsystem 5).

## Subsystem 7

This subsystem also attempts to widen the systems analyst's point of view, in this case by making the previous output be involved in the output at the current stage. Thus,

$$W(t) = k(V(t) + W(t - 1))$$

From this, the analyst learns that she must also broaden her view in the time dimension. This type of delayed action is one of the most difficult things for systems analysts to learn. When we tried to use the function

$$W(t) = k(V(t) + W(t - 2))$$

it seemed impossible for any of the students to discover what was going on. More practice can be used to help develop this atrophied sense of "history."

## Subsystem 8

The W in this subsystem is drawn from a sequence of pseudorandom numbers, but in such a way that slight patterns remain. For instance, the numbers run in odd-even pairs. Here the student is to learn that some systems (although in principle perfectly deterministic) are essentially random given the observer's limited powers and time. What the student should learn to do here is to apply various tests of the hypothesis that she really does not have enough information to solve the problem. A statistical evaluation is of some help here, though it does not, of course, prove anything conclusively. More important is that certain students do *try* to apply statistical tests to this box, while none tried to apply them to Box 6, even though many thought Box 6 was also "random" in some way. Evidently the word "random" is a gloss for at least two quite different perceptions—perceptions we might characterize as "mostly random" (Box 8) and "mostly nonrandom" (Box 6).

# Bibliography

Books that are unpleasant to read give their subjects a bad name. Lord knows, systems analysis and design has a bad enough name already, so I'm not about to recommend any books I didn't enjoy reading.

Titles listed with commentary are recommended as possible next steps after completing this book. The commentary indicates which of many possible directions the step would take. They're only intended to take you a single step. The books themselves contain further references if you want to continue along the same branch.

Other titles are books referenced somewhere in the text. They aren't particularly recommended as next steps from this book, though that doesn't mean they're not recommended as books in general. One or two of them, however, were referenced as bad examples. You'll have to decide which.

ASHBY, W. ROSS. An Introduction to Cybernetics. New York: John Wiley and Sons, 1961.
*Still the best book on the cybernetic view of systems. Some readers will prefer Ashby's* Design for a Brain, *but if you're like me, you'll want to read both. Some mathematics, but nothing that isn't made perfectly clear.*

BOULDING, KENNETH. The Image: Knowledge in Life and Society. Ann Arbor: Ann Arbor Paperbacks, 1961.
*An essential, provocative work for anyone who makes a living working with people's ideas of what constitutes a system. A virtuoso, prize-winning exhibition of what the human mind and pen can do with ideas.*

CHEKHOV, ANTON PAVLOVICH. Lady with Lapdog and Other Stories. Baltimore: Penguin Books, Inc., 1964.

COATS, ROBERT B., AND PARKIN, ANDREW. Computer Models in the Social Sciences. Cambridge, Mass.: Winthrop Publishers, Inc., 1977.
*Simulation is one of the systems analyst/designers most important tools. This is the nicest little book on simulation, in plain practical language.*

DEMARCO, TOM. Structured Analysis and System Specification. New York: Yourdon, Inc., 1978, 1979.

*Tom has the gift of clear writing, backed by the gift of clear thinking and the experience of teaching a tough audience. The best book available on some of the "hard" parts of systems analysis.*

DERRY, T. K., AND WILLIAMS, TREVOR. A Short History of Technology. Oxford: Oxford University Press, 1960 (paperback, 1970).
*Although it contains some inaccuracies, they don't detract from the readability for someone who wants to start getting a historical perspective on the work we do.*

GANE, CHRIS, AND SARSON, TRISH. Structured Systems Analysis: Tools and Techniques. Englewood Cliffs, N.J.: Prentice-Hall, Inc., 1979.
*Gane and Sarson are two leaders in the movement to restructure systems analysis. Their book is worth reading by any systems analyst, beginner or veteran.*

GAUSE, DONALD C., AND WEINBERG, GERALD M. Are Your Lights On? Cambridge, Mass.: Winthrop Publishers, Inc., 1982.
*In this book, Don and I have tried to teach people some things about the art of problem definition. It's light enough to be a catalyst between systems people and normal human beings.*

GILB, TOM, AND WEINBERG, GERALD M. Humanized Input. Cambridge, Mass.: Winthrop Publishers, Inc., 1977.

GRAYBEAL, WAYNE, AND POOCH, UDO W. Simulation: Principles and Methods. Cambridge, Mass.: Winthrop Publishers, Inc., 1980.
*An excellent textbook on simulation, covering the subject rather completely, and in somewhat more mathematical depth than Coats and Parkin.*

LEACH, E. R. Rethinking Anthropology. London: The Athlone Press, 1961, 1966.
*This is the book that inspired the name of the volume at hand. Those readers with some training and interest in anthropology might enjoy studying it for connections between the two works.*

MARCUS, ROBERT. Principles of Specification Design. Robert J. Brady Co., Bowie, MD.
*This is not a book, but a series of animated films accompanied by a worktext. It deals with the problems of obtaining a meeting of the minds between systems analysts and potential computer users. It's particularly useful as an experience to be shared between analysts and their users before they get too far into a new project.*

NAUR, P., AND RANDELL, B. eds. Software Engineering. Brussels: NATO Scientific Affairs Division, 1969.

PARKIN, ANDREW. Systems Analysis. Cambridge, Mass.: Winthrop Publishers, Inc., 1980.
*The best "complete" textbook on systems analysis that I know. Clearly written, clearly illustrated, and well thought out, it's a textbook that can be read with pleasure and profit by any systems analyst, whether green or gray.*

VAN GIGCH, JOHN P. Applied General Systems Theory. New York: Harper and Row, 1978.
*An excellent "textbook" approach to general systems theory, as applied to practical problems. The coverage is very thorough. Though some readers may be put off by the mathematical parts, these are well isolated and not really essential to a useful reading.*

VICKERS, SIR GEOFFREY. Responsibility, Its Sources and Limits. Seaside, Calif.: Intersystems Publications, Inc., 1980.
*The quotations in the preface are from an adaptation of the paper, "Stability and Quality in Human Systems," which is found in this collection of papers by Vickers. Anything Vickers writes is worth reading by the concerned systems person.*

WEINBERG, GERALD M. An Introduction to General Systems Thinking. New York: John Wiley and Sons, 1975.

WEINBERG, GERALD M., AND WEINBERG, DANIELA. On the Design of Stable Systems. New York: John Wiley and Sons, 1979.
*These constitute the first two volumes of a contemplated four-volume series on general systems thinking. I believe they form a good starting place for those whose interest has been aroused about the subject of general systems thinking. Of course, I should think so, as I had a lot to do with writing them, so take my advice with a grain of salt.*

WEINBERG, GERALD M. Learning and Meta-Learning Using a Black Box. Extract from *Cybernetica*, No. 2, 1971. Namur: International Association for Cybernetics, 1971.

WEINBERG, GERALD M. "Stateside," "Phase 2," "From Eagle, Nebraska."
*These columns have appeared in* Datalink *(London),* Computer Careers News *(New York),* Australasian Computerworld *(Sydney), and* bit *(Tokyo) over several years. Many of the essays in this book are adapted from those columns, taking responses from readers into account.*

WEINBERG, GERALD M., YASUKAWA, NORIE, AND MARCUS, ROBERT. Structured Programming in PL/C. New York: John Wiley and Sons, Inc., 1973.

YOURDON, EDWARD. Techniques of Program Structure and Design. Englewood Cliffs, N.J.: Prentice-Hall, Inc., 1975.

# Index